Naval Logistics in Contested Environments

Examination of Stockpiles and Industrial Base Issues

JOSLYN FLEMING, BRADLEY MARTIN, FABIAN VILLALOBOS, EMILY YODER

RAND NATIONAL DEFENSE RESEARCH INSTITUTE

For more information on this publication, visit **www.rand.org/t/RRA1921-1**.

About RAND

The RAND Corporation is a research organization that develops solutions to public policy challenges to help make communities throughout the world safer and more secure, healthier and more prosperous. RAND is nonprofit, nonpartisan, and committed to the public interest. To learn more about RAND, visit www.rand.org.

Research Integrity

Our mission to help improve policy and decisionmaking through research and analysis is enabled through our core values of quality and objectivity and our unwavering commitment to the highest level of integrity and ethical behavior. To help ensure our research and analysis are rigorous, objective, and nonpartisan, we subject our research publications to a robust and exacting quality-assurance process; avoid both the appearance and reality of financial and other conflicts of interest through staff training, project screening, and a policy of mandatory disclosure; and pursue transparency in our research engagements through our commitment to the open publication of our research findings and recommendations, disclosure of the source of funding of published research, and policies to ensure intellectual independence. For more information, visit www.rand.org/about/principles.

RAND's publications do not necessarily reflect the opinions of its research clients and sponsors.

Published by the RAND Corporation, Santa Monica, Calif.
© 2024 RAND Corporation
RAND® is a registered trademark.

Library of Congress Cataloging-in-Publication Data is available for this publication.
ISBN: 978-1-9774-1029-0

Cover: Jerine Lee/ Navy Media Content Services

Limited Print and Electronic Distribution Rights

About This Report

The U.S. Navy has developed concepts for distributed maritime operations (DMO) that use the inherent mobility and firepower characteristics of naval forces to generate multi-axis attacks, complicate enemy targeting, and project power over a wide battlespace. Supporting this concept will require new approaches to logistics and sustainment. Current supply chain processes do not address these requirements adequately. The need for refined understanding extends from the industrial base, where the supply chain originates, all the way to the "last tactical mile" at the end of the distribution chain. This report assesses logistics requirements and identifies improvements, both in terms of capabilities and within supply chain processes, to support DMO in contested environments.

RAND National Security Research Division

This research was sponsored by the Office of the Chief of Naval Operations, Logistics Directorate; and conducted within the Navy and Marine Forces (NMF) Center of the RAND National Security Research Division, which operates the National Defense Research Institute (NDRI), a federally funded research and development center sponsored by the Office of the Secretary of Defense, the Joint Staff, the Unified Combatant Commands, the Navy, the Marine Corps, the defense agencies, and the defense intelligence enterprise.

For more information on the RAND NMF Center, see www.rand.org/nsrd/nmf or contact the director (contact information is provided on the webpage).

Acknowledgments

We appreciate the efforts of our project monitor, CDR Adesina Ekundayo, and we would like to thank our reviewers.

Summary

Issue

To meet the demands posed by today's challenging strategic environment, the Navy has developed its operational concept of distributed maritime operations (DMO). The concept "leverages the principles of distribution, integration, and maneuver to mass overwhelming combat power and effects."[1] This has impacts across the naval force but, specifically of interest to this report, requires an updated resupply and sustainment concept. The Navy logistics community has identified critical challenges that will affect sustainment across the supply chain. They include issues related to acquisition, storage, transportation, and distribution.

However, not all elements of the supply chain are completely understood. Analyses of the enterprise indicate it is not adequately postured to meet the sustainment demands expected under an operational scenario against a near-peer competitor. This report assesses supply chain processes and identifies improvements, both in terms of capabilities and within supply chain processes, to support DMO in contested environments. While we assessed all aspects of the supply chain, we focused on the front end of the acquisition process, specifically the processes and capabilities needed in the production and industrial base for sustainment.

Approach

To assist the Navy in understanding supply chain challenges likely to be faced under DMO in a contested environment, we reviewed Department of Defense (DoD) and Navy policies and regulations related to supply chains, explored relevant military case studies, interviewed key stakeholders involved, and researched academic literature related to our study efforts.

Findings

Our review of Navy supply chain processes identified multiple challenges facing the Navy's ability to support DMO:

- Current supply chain initiatives focus on fixing near-term readiness.
- Current models and demand estimates do not accurately account for DMO requirements. As a consequence, the research team, through its analysis, has estimated the demand.

---

[1] U.S. Department of Defense (DoD), *Advantage at Sea: Prevailing with Integrated All-Domain Naval Power*, December 2020a.

- Misaligned incentives among key stakeholders make it challenging to source adequately to meet wartime demands.
- The Navy is not currently buying to meet demand required for major combat operations, as estimated by the research team, and would be short hundreds of weapons under any operating concept. Even if it could fund these needs, issues in the industrial base capacity would prevent its ability to surge to meet demand.
- Funding mechanisms complicate buying to meet wartime needs.
- Complicated relationships and shared production lines obscure awareness of vulnerabilities in the supply chain.

Mitigation strategies will need to address demand forecasting, budgetary concerns, and industrial base capacity.

Recommendations

To address these challenges, we recommend that the Navy pursue mitigation strategies across three time horizons: near-term (0 to 3 years), mid-term (2 to 7 years), and long-term (5 to 15 years). These strategies are outlined in Table S.1. This approach accepts risk in the near term. If conflict with a near-peer competitor were to break out in the near term, surge capabilities such as the reallocation of munitions and use of emergency mobilization mechanisms would need to be heavily leveraged. But, by accepting risk in the near term, the Navy can prioritize investing in increased inventory across the Future Years Defense Program (FYDP) and making a long-term investment in emerging technology and system design. Across all these time horizons, we recommend the Navy focus efforts on better calculation of demand that uses better engineering models, kill chain system assessments, and live testing.

Table S.1. Mitigation Strategies by Time Horizon

	Force Employment "Near-Term" (0–3 Years) "Surge"	Force Development "Mid-Term" (2–7 Years) "Increase inventory"	Force Design "Long-Term" (5–15 Years) "Long-Term investment"
Munitions	• Reallocate inventory • Add production shifts	• Build factory capacity • Increase inventory through funding	• Adopt modular designs • Use additive manufacturing
Class IX	• Use emergency mobilization mechanisms (e.g., Defense Production Act)	• Use Navy appropriations to fund "kill chain essential" spares	• Rebalance sparing away from legacy aircraft to next-generation platforms (e.g., F-35)
	• Better calculate demand (better engineering models, kill chain basis for system assessment, live testing)		

Contents

Figures and Tables

Figures

Tables

1. Introduction

The U.S. Navy defines distributed maritime operations (DMO) as "an operations concept that leverages the principles of distribution, integration, and maneuver to mass overwhelming combat power and effects."[1] As the Navy evolves toward DMO as a response to the increased capability of near-peer adversaries, it will need to contend with the challenges associated with sustaining distributed units. Supporting this concept will require new approaches to logistics and sustainment. There are several concepts surrounding "sustainment" and "logistics" that can make the discussion confusing. Here we are speaking specifically about the ability to keep combat-ready units forward in contested areas and to ensure their resupply in the event a conflict breaks out in which the supply chains are contested.

Under likely operating scenarios for the Navy against near-peer competitors such as China, the Navy will need to operate in new and different formations. For the past few decades, the United States has assumed that it can resupply forces carrying out high-tempo operations without fear of disruption. However, this is an increasingly untenable assumption as adversaries acquire and develop long-range systems that can readily attack transportation hubs and assembly points. Moreover, as the Joint Force increasingly relies on dispersion and agility, it also creates an increasing challenge for resupplying and sustaining these highly mobile forces.

Current supply chain processes do not adequately address these requirements. Military supply chains originate from where the commodities are produced and end with delivery to the operating units. For this report, a generic supply chain is presented in Figure 1.1. The need for refined understanding extends from the industrial base, where the supply chain originates, all the way to the "last tactical mile" at the end of the distribution chain. If any part of the chain is inadequate, the whole chain will be inadequate. For example, the world has plenty of fuel for effectively every use, but if it cannot be delivered to units in combat, sustainment is insufficient. In some cases, the issue may not be delivery but the inability to produce the commodities in sufficient numbers quickly enough to meet the demand generated in battle.

Figure 1.1. Generic Military Supply Chain

Acquisition ▶ Storage ▶ Transportation ▶ Distribution

---

[1] DoD, 2020a.

There are numerous ways that forces operating in dispersed and contested environments could be sustained. Some key elements of sustainment, particularly personnel and reusable items, should be organic to the force itself. Wise use of prepositioning may reduce the amount of material that requires delivery. However, many (if not most) items, such as fuel and munitions and some spare parts, will need to be delivered.

These sustainment challenges apply across all commodities. Some challenges, such as provision of fuel, revolve around distribution and might not reach deep into national supply chains. Two commodity classes, however, stretch across the spectrum: munitions and spare parts. Like fuel, these must be distributed to widely dispersed units. However, unlike fuel, the worldwide supply is not plentiful, with all ordnance and most spare parts requiring manufacture in the United States, using the existing U.S. defense industrial base. Therefore, we decided to focus our research efforts on ordnance and spare parts because they provide useful case studies that help illuminate supply challenges facing the Navy, particularly farther forward in the early stages of the supply "kill chain."

Research Objective and Approach

The objective of this report is to assess supply chain processes and identify improvements to support DMO in contested environments. We focused on early stages of the acquisition process because analyses identified a significant gap in understanding the requirements that may need to be met in a demanding operational scenario and the ability of the industrial base, the Navy, the Defense Logistics Agency (DLA), and other critical stakeholders to meet the expected demand. We further refined the project scope to focus on two specific commodities (munitions and naval aviation spare parts) that illuminate these challenges. Our intent was to answer several questions relating to the degree of vulnerability resulting from supply chain shortfalls in the projected operational environment and some of the options available to meet these vulnerabilities.

There are some key differences between munitions and spare parts to consider. Demand for munitions is only a factor during actual periods of warfare, and while the expected demand is predictable, it does not routinely occur. Demand for spare parts, however, may vary considerably depending on the operating environment. In addition, munitions are purchased directly by appropriated funds; spare parts are in general purchased through DLA working capital fund (WCF) mechanisms.

Figure 1.2 depicts the challenges the Navy faces in dealing with the resupply in commodities that might only be demanded by the military and perhaps demanded only by the military during war. Replacement parts that experience high-demand levels during peacetime operations will likely be easily replaced because the supply chain is robust and understood. Critical parts and munitions that are used less frequently in normal operations but are essential while carrying out combat will have inventory management challenges. We will discuss these challenges in more

Figure 1.2. Essentiality Versus Ease of Procurement

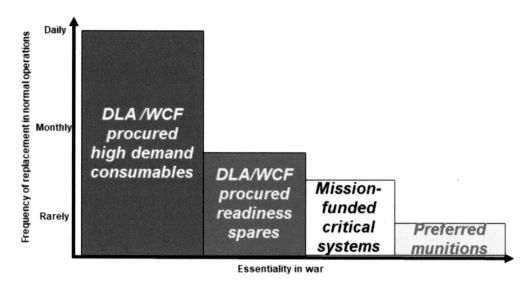

detail, but for now, the issue to consider is that normally consumed commodities will continue to be readily available while critical commodities likely will not be.

Our approach included a mixture of qualitative methods. We reviewed DoD and Navy policies and regulations to identify current naval supply chain processes and explored relevant case studies to provide context for those findings. Interviews with stakeholders further identified current processes, challenges, and mitigation options. We conducted limited data analysis where we had access to material or where material had been collected. However, the limited amount of available data stands as a critical shortfall. Finally, our review of academic literature helped us determine best practices.

Outline of This Report

In this report we focus on two specific naval supply chains—critical munitions and naval aviation repair parts. Chapter 2 outlines the current munitions supply chain and the challenges it faces. Mitigation options to address those challenges are presented in Chapter 3. Chapter 4 identifies the current state of naval aviation supply chains and their associated challenges. Chapter 5 presents possible improvements. Chapter 6 summarizes our overall findings and recommendations based on our exploration and analysis of the two supply chains and presents areas for further research.

2. Current State of Munitions Supply Chains

Munitions are unique to military supply chains. Other nonmilitary organizations need fuel and spare parts, but only the military uses high-explosive, long-range weapons in any quantity. However, within the military, munitions use has unique characteristics that differentiate it from other classes of supply. Most notably, munitions are not used in normal peacetime operations unless required for training. Therefore, the military only expends small numbers of advanced weapons during peacetime but can expect to expend hundreds or thousands in wartime.

The challenges of building and maintaining inventory are extensive across the supply chain. They include low production rates that fail to meet the expected demand required to support major combat operations—including the specific operations contemplated in DMO—and missile component production constraints (limited suppliers, long lead times, foreign dependencies). In this chapter we examine these challenges in more detail.

Deriving Demand for Munitions from War Plans and Scenarios

Actual weapons usage in wartime may differ from expectations as operational and tactical situations evolve. But, for the contested environments directed for planning in the National Defense Strategy of 2022, the demands are generally very clear and can be discussed in general terms in an unclassified setting.

For purposes of this report, we are considering two naval munitions: the long-range anti-ship missile (LRASM), otherwise known as the AGM-158C, which has a common rocket body with the joint air-surface standoff missile (JASSM); and the Maritime Strike Tomahawk missile (MST), otherwise known as the Block Va variant.[1] These munitions are specifically intended to be delivered as standoff weapons against surface ships, and the required inventory levels can be reasonably estimated from—simply—the number of ships the Navy needs sink, given an expected level of air defense. There are no suitable close-in substitutes, and there is no other target set that these munitions might reasonably service. Munitions demand may not always be so readily specified, but for these munitions, the demand can be reasonably specified given some relatively well-understood scenarios.

To estimate the demand from a Western Pacific scenario and the ability of the industrial base to meet this demand, the following assumptions were made based on force structure, expected

---

[1] There are three variants of the Tomahawk Block V munition: (1) The Block V will simply receive upgraded NAV/COMMs packages; (2) the Block Va, also known as the MST, will be able to target ships; and (3) the Block Vb will retain anti-ship capability and be upgraded to better target hardened and buried land targets with a joint multiple effects warhead system. See Richard Scott, "Cruise Control, Block V Missile Starts a New Chapter for Tomahawk Line," *Jane's International Defence Review*, December 9, 2021a.

targets in the most salient scenarios, the weapons load-outs of the platforms, and unclassified descriptions of tactics and firing doctrines:

- Approximately 20 U.S. Navy (USN) ships (roughly equivalent to two carrier-strike groups) with 30 missiles each would be able to provide 600 missiles.
- Approximately 80 People's Liberation Army (PLA) naval ships would each require four missiles to sink. To accommodate any losses of munitions due to intercept or electronic countermeasures, however, a multiplication factor of four was used to assume a total of 1,280 missiles needed.
- Current production and recertification rates hold constant for the LRASM and Block V upgrades of Tomahawk missiles, respectively.

Table 2.1 shows the estimated demand for anti-ship munitions for a Western Pacific scenario. Over the course of a 65-day conflict, a total of 800 MSTs and 1,200 LRASMs are needed to engage with a PLA naval fleet. Half of the munitions (400 MSTs and 600 LRASMs) are assumed to be expended within the first 15 days of conflict. The other half is assumed to be expended over the course of day 45 to day 60 after conflict begins.

Table 2.1. Estimated Demand for Munitions from a Western Pacific Scenario

Munitions	Days 1–15	Days 45–65	Total
Tomahawk—MST (Block V and later)	400	400	800
LRASM/AGM-158C	600	600	1,200

SOURCE: RAND Analysis.
NOTE: The range of days refers to the dates after conflict begins.

Munition Procurement and Inventory

As shown in Table 2.2 and Table 2.3, the USN has aimed to procure both the LRASM and Tactical Tomahawk (TACTOM) Block V variant munitions in the past and in future fiscal years. Lot 15 was the last of Block IV TACTOMs delivered in 2020.[2] Lot 16 and later are to be built to the Block V standard, which would include a navigation/communications (NAV/COMMs) package upgrade but will not have the capability to target maritime vessels.[3]

Based on procurement rates up to fiscal year (FY) 2021, there are about 147 LRASMs currently in the U.S. inventory, with plans to procure an additional 162 LRASMs by 2025 (Table 2.2).

---

[2] Scott, 2021a.

[3] Scott, 2021a.

Table 2.2. USN Munition LRASM Procurement Rates

Munition	Previous	FY19	FY20	FY21	FY22	FY23	FY24	FY25	Total
LRASM/AGM-148C	49	33	17	48	48	48	48	18	**309**

SOURCE: U.S. Department of Defense, *Fiscal Year (FY) 2021 Budget Estimates: Navy Justification Book, Volume 1 of 1, Weapons Procurement, Navy*, February 2020b.

Table 2.3. USN Munition Tomahawk Procurement Rates

Munition	FY16	FY17	FY18	FY19	FY20	FY21	Total
TACTOM Block V	N/A	N/A	N/A	N/A	90	155	**245**
TACTOM Block IV	149	196	44	56	N/A	N/A	**445**

SOURCE: DoD, 2020b.
NOTE: The annual production minimum sustainment rate for the TACTOM Block V is 90 missiles.

In addition, the USN has begun upgrading all extant Block IV Tomahawks into the Block V variant with a goal to upgrade approximately 3,992 missiles. Table 2.4 shows the delivered munitions and future plans for the Block V variant.

Table 2.4. Tomahawk Block V Recertification Rates

Fiscal Year	Funding	Upgraded Thus Far	Remainder
FY21	112	80	3,912
FY22	156	Unknown	<3,912
FY23*	300	Unknown	<3,912

SOURCE: Richard Scott, "Cruise Control, Block V Missile Starts a New Chapter for Tomahawk Line," *Jane's International Defence Review*, December 9, 2021a.
NOTE: * The 2023 figures are assumed based on 300 NAV/COMM upgrade kits to be procured in FY 2022. Although FY 2021 included funding for 112 Block V upgrades, only 80 were delivered.

One of the Block V variants, the Block Va MST, will be able to target moving maritime targets to provide an anti-ship capability. The MST is not scheduled to attain initial operating capability (IOC) until 2024.[4] We were unable to confirm what percentage of future Block IV upgrades would be dedicated to the Block Va MST variant once operational. Thus far, a total of

---

[4] Scott, 2021a.

116 Block Va MST kits have been included or are proposed in budgets from FY 2021 to FY 2023.[5]

Finally, recertification will add 15 years of service life to the converted Block IV Tomahawks.[6] Given their 30-year lifetime, new Block V production (Lot 16 and later)[7] could keep Tomahawks in USN inventory into 2050 and beyond, with servicing needed midway or about 2035.

Current U.S. Navy Inventory

Table 2.5. FY 2021 USN Inventory for Anti-Ship Munitions

Munition	Inventory
TACTOM Block V	245
TACTOM Block IV	~2,900
LRASM/AGM-158C	147

SOURCE: Tomahawk Block V and LRASM numbers were taken from DoD, 2020b. Tomahawk Block IV numbers were taken from Janes, "Weapons: Naval, Missiles, United States," updated April 7, 2021b.

Demand Outside the U.S. Navy

Long-range standoff munitions are desirable candidates to reach PLA naval ships without significant risk of losing aircraft. This makes their use by both the USN and the U.S. Air Force (USAF) all the more likely. Furthermore, the long-range standoff munitions that are being considered for the anti-ship mission share similarities in their design; the JASSM-extended range (JASSM-ER) and LRASM munitions share components and would hence share lower-tier suppliers of these components. "Both JASSM-ER and LRASM missiles are being produced on the same line at Lockheed Martin Missiles and Fire Control's facility in Troy, Alabama, using common tooling and production processes. The USN and USAF jointly procure the AGM-158C via an air force contract that aligns with AGM-158B JASSM-ER procurements for the air force."[8]

Were a conflict to begin, and if orders for both munitions were to be contracted for resupply, it could result in second-tier suppliers (and below) having to prioritize components for one munition over the other. Moreover, it is entirely possible one service would need to be prioritized

---

[5] The budgeted quantities for MST kits for FY21 and FY22 are 15 and 29, respectively. The proposed FY23 MST kit quantity is 72. See Richard Scott, "Anti-Surface Warfare: USN Plans Funding Towards a Bigger Punch, Longer Reach," *Jane's International Defence Review*, April 29, 2022.

[6] Scott, 2021a.

[7] Scott, 2021a.

[8] Scott, 2022.

over the other should both the USN and USAF require resupply, which is very likely. It is therefore prudent that considerations be made to account for demand outside the USN.

The USAF has procured and has planned to further procure JASSM-ER and LRASM missiles. Table 2.6 shows the quantities the USAF has procured and plans to procure from FY 2021 to FY 2025.

Table 2.6. USAF Demand for Long-Range Standoff Weapons to 2025

Munition	Procured	Planned by 2025	Remainder	FY22	FY23	FY24	FY25
JASSM-ER/AGM-158B	4,444	7,200	2,756	512	516	520	509
LRASM/AGM-158C	0	410	410	N/A	44	52	28

SOURCE: Planned USAF numbers for JASSM/JASSM-ER are taken from Mark A. Gunzinger, "Affordable Mass: The Need for a Cost-Effective PGM Mix for Great Power Conflict," *Mitchell Institute Policy Paper* 31, November 2021; procured and planned procurement numbers for FY 2022 to FY 2025 for JASSM/JASSM-ER and LRASM are taken from U.S. Department of the Air Force, *Department of Defense Fiscal Year (FY) 2021 Budget Estimates: Air Force Justification Book, Volume 1, Missile Procurement*, Air Force, February 2020, pp. 19 and 22.

The USAF has planned to procure a total of 7,200 JASSM-ER/AGM-158D missiles by 2025. So far, it has received 4,444 of them, leaving a remainder of 2,756 missiles yet to be delivered. A total of 2,057 JASSM-ERs are budgeted for over the next four fiscal years, leaving 699 unaccounted for.

The USAF also plans to procure 410 LRASMs by 2025, but none have been delivered thus far, and only 124 have been budgeted for over the next four fiscal years, leaving 286 missiles unaccounted for.

In total, then, the USAF plans to procure 3,166 munitions (JASSM-ERs and LRASMs) using the same production line and that need to be delivered by 2025, with only 2,181 budgeted and 985 unaccounted for. Whether or not these quantities will increase in future budgets is uncertain and adds to factors that may constrict the industrial base's ability to deliver munitions to the USN.

Assuming the current budget quantities stay constant over the next four fiscal years, the production rate for a combined JASSM-ER and LRASM production line averages 545 munitions per year.

Can the Defense Industrial Base Support a Western Pacific Scenario?

Tomahawk

Current recertification rates are aimed at converting Block IV TACTOMs to the Block V variant. New production lot numbers (16 and later) are also aimed at producing the Block V variant. This means none of the new Tomahawks will have the required capability to target surface vessels. Additionally, the MST is not scheduled to attain IOC until 2024. If current budgeted quantities remain consistent, it is likely that the USN will only have 116 MSTs in its inventory by 2025 and will not reach the estimated demand for a Western Pacific scenario (800 munitions).

If recertification rates remain constant, it will take approximately 13 years to fully upgrade Block IV's to the Block V, Block Va, or Block Vb variants.[9] If Block Va MSTs reach IOC in 2024, *and* if recertification of Block IV TACTOMs upgrades to the Block Va MST instead of the Block V, it will deliver an additional 600 munitions by 2025. If, in addition, all new production lots output the Block Va variant instead of the Block V at a rate of 155 munitions per year (the budgeted quantity for FY 2021) for the next three years, an additional 465 munitions may enter the USN's inventory by 2025 for a total of 1,181 Block Va MSTs. However, the likelihood of these two courses of action happening is uncertain.

Long-Range Anti-Ship Missiles

The average production rate of JASSM-ER has been 257 units per year, but based on Table 2.5, the production rate will increase to approximately 500 or more units per year starting in FY 2022 to FY 2025. The USAF plans to procure 28 to 52 LRASMs per year (Table 2.6) over the same period while the USN plans to procure between 18 and 48 LRASMs each year (Table 2.2). The combined quantities on the shared production line across the services averages 585 munitions per year.

As stated above, total USAF demand for standoff munitions is 3,166. Based on our assumption of 1,200 LRASMs needed by the USN for anti-ship targeting and fires, and with 147 currently in the USN's inventory (Table 2.5), the total demand from both services is 4,219 munitions. Given the current production rates, it could take 5.4 years to 12.3 years to produce USAF precision-guided missiles (PGMs) alone, and 7.2 years to 16.4 years to produce all 4,219 munitions demanded by the two services.[10]

Given current budgeted quantities for LRASMs for the USN for FY 2022 to FY 2025, and the 147 LRASMs in the current inventory, the USN will only be able to inventory 309 LRASMs by 2025, well short of the estimated 1,200 needed for a Western Pacific scenario.

Moreover, when we consider the potential manufacturing constraints within the munitions supply chain, it is difficult to imagine either Tomahawks or LRASMs meeting estimated demand by 2025. To elaborate, we turn our attention to these constraints.

---

[9] Approximately 3,912 Block IV TACTOMs will need upgrades and recertification. At a rate of 300 units per year (the procurement quantity estimated for FY23), it will take approximately 13.0 years to complete the Block V conversion.

[10] Total USAF demand is 3,166 JASSM/JASSM-ERs and LRASMs. This 3,166 divided by the previous production rate of 257 equals 12.3 years. When divided by the projected production rate of 585, the estimate yields 5.4 years. Similar arithmetic yields 16.4 years and 7.2 years for 4,219 munitions when the demand from the two services are combined.

Supply Chain and Manufacturing Constraints

Every manufacturer has constraints within its supply chain. However, when that manufacturer is a supplier to the DoD, these simple constraints can affect readiness to create a preparedness deficit in the future. It is judicious to keep these constraints in mind when considering the need for resupply during conflict. Several studies review the constraints and limitations of the solid rocket motor supply chain, but they are broadly applicable to the munitions supply chain and defense industrial base as a whole.[11] Table 2.7 lists some well-known constraints within the munition component supply chains based on these studies.

Table 2.7. General Manufacturing Constraints Across Munitions

Missile Components	Subcomponents or Material Inputs	Constraints
Encasement	• Composites and metals	• Limited suppliers
Payload	• Ordinance	• Highly regulated—safety, environmental
Communications	• Radio frequency (RF) receiver/transmitter	• Supply chain shortage—microchips
Control	• Actuator	• Foreign dependencies—rare earth magnets
Guidance and navigation	• Seeker, dome, radar, Global Positioning System (GPS)	• Supply chain shortage—microchips
Power	• Lithium-ion batteries	• Foreign dependencies—raw materials and battery cells
Propulsion	• Turbofan	• Complex manufacturing processes—fan blades Limited suppliers—high-temperature metals
	• Solid rocket motors (SRM)	• Limited suppliers—two SRM original equipment manufacturers (OEMs) Single-source suppliers—propellant materials Long-lead items—nozzles, cones, throat Highly regulated—safety, environmental
Airframe	• Composites and metals	• Limited suppliers

SOURCE: RAND Analysis.

---

[11] See U.S. Department of Defense, *SRM Industrial Capabilities Report to Congress: Redacted Version*, June 2009; Government Accountability Office, "Solid Rocket Motors: DOD and Industry Are Addressing Challenges to Minimize Supply Concerns," GAO-18-45, October 2017; Brian Gladstone, Brandon Gould, and Prashant Patel, "Evaluating Solid Rocket Motor Industrial Base Consolidation Scenarios," IDA Research Notes, Spring 2016; Thomas A. Ganey, *Endangered Species-Solid Rocket Motor Manufacturers: Preventing a National Asset Extinction*, Air University, Air Command and Staff College, Maxwell Air Force Base, February 11, 2011; U.S. Department of Commerce, Bureau of Industry and Security Office of Technology Evaluation, *U.S. Rocket Propulsion Industrial Base Assessment*, 2018; and Theresa S. Hull, "Audit of Purchases of Ammonium Perchlorate Through Subcontracts with a Single Department of Defense–Approved Domestic Supplier," Office of the Inspector General of the Department of Defense, July 9, 2020.

In addition to these constraints, there are constraints inherent in any manufacturing process, such as availability of equipment and tooling and limited material inputs.

LRASM Constraints

Because of the LRASM's shared design with the JASSM-ER, and the JASSM-ER's commonalities with the AGM-158A JASSM, the LRASM also shares a production line, components, and suppliers with these other missile systems. Should a surge in demand occur for both munitions at the same time (e.g., during conflict), there could be delays to the delivery of munitions for resupply. Table 2.8 highlights some of the commonalities between the munitions.

Table 2.8. Shared Components Between LRASM, JASSM-ER, and JASSM

Component	LRASM	JASSM-ER	JASSM
FMU-156/B ESAF fuze*	X	X	X
WDU-42/B (J-1000) penetration and blast fragmentation warhead	X	–	X
F107-WR-105 turbofan engine	X	X	–
Enhanced digital anti-jam GPS receiver	X	X	–
Vertical tailfin	X	X	–

SOURCES: Janes, "AGM-158C Long-Range Anti-Ship Missile (LRASM) Weapons: Air-Launched," December 15, 2021a; Richard Scott, "USN Axes JSOW ER in Favour of JASSM-ER Buy," Jane's Missile and Rockets, June 9, 2021b.
NOTE: * Another fuze is in development, the FMU-162/B EASF, which would replace the FMU-156/B ESAF in the JASSM, JASSM-ER, and presumably, the LRASM.

Additional Constraints

In addition to the general constraints detailed above, there are a few more that warrant attention:

- The LRASM[12] and Tomahawk[13] share a supplier for the turbofans—Williams International—and since the LRASM shares a turbofan with the JASSM-ER, all three munitions share the same supplier.
- LRASM and JASSM-ER share a production line at Lockheed Martin Missiles and Fire Controls.[14]
- All new Tomahawks are produced at a single Raytheon facility.[15]

---

[12] Janes April 7, 2021.

[13] Scott, 2021a.

[14] Scott, 2022.

[15] CNN, "Inside a Tomahawk Missile Factory," video, YouTube, undated.

Case Studies

To demonstrate how actual weapon usage in wartime may differ from expectations, we highlight a few case studies from the past 30 years of U.S. operations where munitions were strained and called for a variety of mitigation strategies. Additionally, these case studies highlight specific applications and potentially some elements required for the success of mitigation strategies in a diverse framework of conflict scenarios and munition demands. The cases include Operation Desert Storm, Operation Allied Force, Operation Enduring Freedom, and Operation Inherent Resolve.

Operation Desert Storm (1991)

The U.S. mission to remove Iraqi forces from Kuwait, known as Operation Desert Storm, was characterized by unpredicted demands for a variety of wartime materials, including munitions. By the final weeks of conflict, news outlets reported that there was a strain on munitions, with some "dumb" munition stocks depleted below a 10-day supply, and that the United States had requested NATO assistance to manage supply.[16]

While the conflict resolved a few weeks after reports of depleted stocks, the munitions stock challenges present a case where munitions were available at the outset of operations but needed to be reallocated to meet shifting demands throughout the conflict. Conflict is guaranteed to be riddled with uncertainty, but Operation Desert Storm presented unique and unexpected demand challenges as the conflict developed. The USAF had the appropriate quantity and type of munitions to meet the operational demands of the conflict but had to quickly transport these munitions to different locations. RAND research assessed that much of the success of the rapid relocations is attributed to Saudi Arabia's well-developed infrastructure.[17] At the end of the conflict, it is reported that the USAF dropped 69,000 tons of munitions.[18]

In Operation Desert Storm's success, this case provides an opportunity to consider different elements that came together successfully in the conflict and how they may change in different scenarios. For example, if the conflict did not resolve when it did and lasted months or perhaps years longer, how would demand and U.S. munitions use change to adapt? If Saudi Arabia did not have the infrastructure to support the rapid movement of munitions, what opportunities would the United States have to access key munitions, or what mitigations would have been used? Operation Desert Storm highlighted not only the role of munition quantities and unexpected

---

[16] Knight-Ridder, "Allies Reportedly Facing Ammunition Shortage. Some Fear Bullets in Short Supply for a Ground War. Persian Gulf Showdown," *Baltimore Sun*, February 13, 1991.

[17] Raymond A. Pyles and Hyman L. Shulman, *United States Air Force Fighter Support in Operation Desert Storm*, Santa Monica, Calif.: RAND Corporation, MR-468-AF,1995.

[18] Pyles and Shulman, 1995.

demands in conflict but also the significance of logistics capacity and the movement of munitions to meet demand.

Operation Allied Force (1999)

In 1999, NATO carried out a 78-day air campaign against the Serbian military and the Slobodan Milošević regime for their occupation of Kosovo and ethnic cleansing of Albanians in the region. During NATO operations, U.S. forces went through scarce and expensive consumables at an unprecedented rate.[19] By the end of the first week of the conflict, the USAF stocks of conventional air-launched cruise missiles (CALCMs) had fewer than 100 missiles remaining.[20] In a DoD report to Congress, the USAF proposed converting nuclear-configured air-launched cruise missiles (ALCMs), though the process would take more than a year.[21] Tomahawk land-attack cruise missiles (TLAMs) were also used heavily in the outset of the conflict, as the Navy reportedly had an estimated 2,500 remaining in stock after the first week of the conflict, with the Navy stating that it would have to "manage the inventory very carefully."[22] The U.S. Navy requested $431 million in emergency funding to convert 624 Tomahawk cruise missiles to the land-attack model.[23] Additionally, the U.S. mobilized joint direct attack munition (JDAM) in the conflict, despite it being in the testing phase at the time of commitment. Between 651 to 656 JDAMs were used over the course of the conflict, and the DoD requested funding to acquire 11,000 additional kits.[24] To meet the increase in demand, Boeing only added two workers to the JDAM production line.[25] The Lot 2 order for JDAMs increased from a planned 180 kits per month to 410 (peaking at 450) and took about four months to reach the new production rate. Boeing credited a tightly integrated supplier base as a critical enabler of this feat.[26]

While the NATO bombing campaign only spanned slightly over two months, the implications of perceived munition shortages led the DoD to consider different applications of munitions, emergency funding, and alternative mitigation strategies, especially in the case of long-term conflict. In the face of looming shortages, the DoD established a task force to reallocate

---

[19] Benjamin S. Lambeth, *NATO's Air War in Kosovo: A Strategic and Operational Assessment*, Santa Monica, Calif.: RAND Corporation, MR-1365, 2001.

[20] Chris Plante and Charles Bierbauer, "Pentagon's Supply of Favorite Weapon May Be Dwindling," *CNN*, March 30, 1999.

[21] U.S. Department of Defense, *Report to Congress: Kosovo/Operation Allied Force After-Action Report*, January 31, 2000.

[22] Plante and Bierbauer, 1999.

[23] DoD, 2000.

[24] DoD, 2000.

[25] Frank Wolfe, "Pentagon Speeds Up JDAM Delivery for Possible Kosovo Use," *Defense Daily*, Vol. 202, No. 12, April 16, 1999.

[26] Charles H. Davis, *The JDAM Experience and DPAS*, The Boeing Company, briefing, undated.

industrial resources as needed for munitions with Department of Commerce assistance.[27] The task force was intended to determine key weapons systems in the conflict and direct the supply chain, and common components, to fulfill priority demands. As this conflict was relatively short, this task force only had limited involvement across the supply chain. Ultimately, the task force was designed to direct priorities for conflicts that would be drawn out over at least a year.

NATO operations in Kosovo demonstrate how an intense conflict, or the rapid use of preferred munitions, can create pressures and the need for mitigation strategies in the early weeks of a conflict. There is a clear dissonance between delivering a decisive end to conflict quickly by using preferred munitions and conducting large-scale operations that require significant munitions supply versus maintaining sufficient stock of those preferred munitions to meet the demands of the conflict—however long it may turn out to be. Conflict is complex and unclear, and it is difficult to definitively know when a conflict will end, so balancing the demands of the conflict as it evolves and maintaining preparedness of the uncertain future is difficult. In this case, the DoD applied mitigation strategies of mobilizing new capabilities such as JDAMs, converting not-preferred munitions to meet the demands of the conflict (as seen with a request to convert nuclear-tipped ALCMs to CALCMs and convert more Tomahawks to the latest model), and establishing a process to redirect the supply chain to address demands based on conflict demand and priority. The campaign ended before many of these initiatives were completed, but the case demonstrates the trade-offs of brief campaigns and the uncertainty of a protracted conflict.

Operation Enduring Freedom (2001)

Operation Enduring Freedom, informally called the Global War on Terrorism, began with the series of terrorist attacks on U.S. defense and commercial centers on September 11, 2001, and included an extensive demand on resources over the next 13 years, with the operation's end on December 31, 2014. The United States carried out extensive military campaigns against al Qaeda in Afghanistan and the Taliban and drew on U.S. smart munitions—leading to a risk of shortages over the course of the operation.

During Operation Enduring Freedom, the USN and USAF used more smart bombs than anticipated, resulting in shortages in-theater. In February 2002, shortly after the beginning of the operation, Gen Richard Myers, the former vice chairman of the Joint Chiefs of Staff, reported that 60 percent of weapons dropped were smart bombs.[28] During the conflict, the Navy reported

---

[27] Anthony H. Cordesman, *The Lessons and Non-Lessons of the Air and Missile Campaign in Kosovo*, Washington, D.C.: Center for Strategic and International Studies, revised August 2000.

[28] U.S. Senate, Committee on Armed Services, "Conduct of Operation Enduring Freedom," Hearing Before the Committee on Armed Services United States Senate, 107th Cong., 2nd Sess., February 7 and July 31, 2002.

that they "nearly ran out of JDAMs," to which the Air Force helped meet Navy demands.[29] In addition to intra-DoD support to meet munition demands, the DoD moved 16,000 tons of munitions "borrowed" from other areas of responsibility (AORs) using both air and naval assets.[30] The majority of the smart bombs expended during this conflict were JDAMs, with an estimated 4,500 dropped in Afghanistan over the duration of the operation.[31]

This case study highlights how quickly a long-term mission, in this case 13 years, can place strain on munitions and how mitigation strategies may create stress on other assets to meet the munition demands in a timely and sufficient manner. In this operation, the USAF and USN had to dedicate assets to move munitions into the theater, which not only left the other AORs at risk of shortage but required dedicated airlift and sealift assets to transport the munitions that may have otherwise been dedicated to other missions if munition demand was met in-theater. Munitions, and shortages of munitions, in a conflict can affect the utilization of a broad range of assets, as seen with airlift and sealift in this operation.

Operation Inherent Resolve (2014)

Operation Inherent Resolve began in 2014 as a mission to counter the Islamic State of Iraq and Syria (ISIS) with military action in Iraq and Syria and related missions in Libya. During this operation, an unusually large number of missiles were deployed in strikes against ISIS, resulting in shortages that lasted until 2020.

As of 2017, the DoD reported 13,331 strikes in Iraq and 11,235 in Syria by the coalition—a total of 84 countries involved in the operation.[32] To mitigate some of the shortages, the DoD reallocated munitions from other AORs to meet mission demand.[33] Additionally, Lockheed Martin and Boeing, two key U.S. munitions manufacturers, took different approaches to increasing production. Lockheed Martin expanded manufacturing capacity while Boeing increased production by adding production shifts.[34] Increased demand for missiles, particularly JDAM and AGM-114 Hellfire, an air-to-ground munition, has persisted since 2015. The U.S. Army granted $18 million to Lockheed to increase production of the Hellfire from 500 to

---

[29] Benjamin S. Lambeth, *Air Power Against Terror: America's Conduct of Operation Enduring Freedom*, Santa Monica, Calif.: RAND Corporation, MG-166-1-CENTAF, 2006.

[30] Robert S. Tripp, Kristin F. Lynch, John G. Drew, and Edward W. Chan, *Supporting Air and Space Expeditionary Forces: Lessons from Operation Enduring Freedom*, Santa Monica, Calif.: RAND Corporation, MR-1819-AF, 2004.

[31] "Joint Direct Attack Munition," *Defense Daily*, 2022.

[32] U.S. Department of Defense, "Operation Inherent Resolve: Targeted Operations to Defeat ISIS," 2022.

[33] Paul D. Shinkman, "ISIS War Drains U.S. Bomb Supply," *U.S. News*, February 17, 2017.

[34] Marcus Weisgerber, "Bombs Away! Lockheed Expanding Missile Factories, Quadruples Bomb Production for ISIS Long Haul," *Defense One*, March 16, 2016.

650 missiles per month.[35] In the FY 2017 budget, 45,000 new smart bombs were requested to address the demands of the operation.[36] Since late 2021, Operation Inherent Resolve has transitioned to a noncombat mission to "advise, assist, and enable Iraqi forces."[37]

In Operation Inherent Resolve, U.S. forces faced demand for specific munitions to meet the demands of the mission, and much of the demand of this mission required JDAM and Hellfire missiles. Additionally, the DoD leveraged multiple different mitigation strategies to ensure demand was met, including reallocating munitions from other AORs and then requesting increased production across the supply chains. Strategies to meet demand also varied from different suppliers, as Lockheed Martin and Boeing increased production, but through different means. Finally, it is important to highlight Operation Inherent Resolve as another case study where the tempo of conflict, especially in the opening weeks and months, quickly placed pressure on existing stocks of munitions viewed as essential to the conflict. Unlike some of the other case studies presented, Operation Inherent Resolve lasted for years and therefore required longer-term mitigation strategies beyond reallocation to meet the demands of the conflict.

Case Study Conclusions

Many of these cases faced unexpected demand for specific munitions that can occur at the outset or mid-conflict. Some conflicts faced munitions shortages at the outset while others faced shortages because of shifting strategy and policy over the course of the conflict.

Shortages can occur at any point in the supply chain. Some of the cases experienced shortages in-theater that required allocation or borrowing from other AORs; others required increased manufacturing. In part, the mitigation strategies reflected in these case studies demonstrate the nature of some strategies as short-term mitigation, such as borrowing from other AORs, while some mitigations require long-term application, such as increasing production at different parts of the supply chain or converting existing munitions to those in demand. Additional research could pursue how these shortages may compound one another—for example, how much does "borrowing" from other theaters generate shortages, and consequently risk, elsewhere?

Summary

Analyses of munitions supply chains identify that production rates and inventory levels for critical munitions required to execute operations against a near-peer competitor in a DMO environment are critically low. Numerous production and supply chain constraints prevent the

---

[35] Jared Keller, "The Pentagon Has Dropped So Many Bombs on ISIS We're Literally Running Out," *Task and Purpose*, May 1, 2017.

[36] Lara Seligman, "Air Force Wants Smart Bomb Increase for ISIS Fight," *Defense News*, April 1, 2016.

[37] Staff Report Joint Operations Command, "Combat Role in Iraq Complete; Invitation from Iraq Reaffirmed to Advise, Assist, Enable," Operation Inherent Resolve, December 9, 2021.

Navy and the industrial base from increasing demand to meet the number of munitions required. Additional complications include the shared production line with other services, which would make it difficult for both services to be prepared for conflict at the same time. Case studies identify a need to pursue both near-term and longer-term mitigation strategies to overcome these challenges.

3. Munitions Supply Chain Mitigation Strategies

To resolve the challenges outlined in Chapter 2, the Navy will need to pursue mitigation strategies over near-, mid-, and long-term planning horizons. Most mitigations primarily revolve around adding supply or improving the ability to move inventories to the point of need. Demand-suppression mitigation strategies are not useful in this case: Munitions are combat essential, and because of the likely scenarios expected when facing a near-peer competitor, substitutes are not a suitable strategy. Therefore, in this chapter we identify possible mitigation strategies that are applied to different planning horizons. These time horizons include near-term force employment (0 to 3 years), mid-term force development (2 to 7 years), and long-term force design (5 to 15 years). It should be noted that many of the mitigation strategies in the near-term force employment time frame accept a level of risk, as the most beneficial strategies to address challenges will not be able to be implemented until farther out because of budgetary processes.

Force Employment Short-Term Strategies (0–3 Years)

In the immediate future, there are significant constraints on mitigation strategies that can be applied on the munitions supply chain to meet surge demand. Short-term strategies work within the framework of munitions that are currently being produced or are about to reach IOC,[1] fixed funding for munitions and the targeted supply chains, and a fixed force design and program of record for applicable munitions. These short-term strategies are typically intended to be applied quickly and to meet surge demands quickly, but they have limited long-term applicability and sustainability. These mitigation strategies include reallocating inventory, expanding industrial capacity, and adding production shifts.

Reallocate Inventory

Reallocation of inventory is simply moving munitions from one location to another, whether within the theater or from another combatant command (CCMD), to meet the demand as it arises. This strategy is not new and has been used previously (Operation Enduring Freedom) to address unanticipated demand and changing strategy.[2]

---

[1] IOC is defined as being attained when "units and/or organizations in the force structure scheduled to receive a system have received it and have the ability to employ and maintain it. The specifics for any particular system IOC are defined in that system's Capability Development Document" (Defense Acquisition University). Consideration of IOC includes the number of assets required, the activities (such as training), and operational units that will employ the capability.

[2] Discussion with Naval Supply Systems Command (NAVSUP) Ammunition Logistics Center official, November 2021.

One of the key benefits of this strategy is that it is most likely the fastest, cheapest, and simplest method of increasing inventory in one location—which makes it ideal as a short-term strategy. The key expenses of this strategy are the transportation and logistics investment required to move the allocated inventory to meet demands in a timely manner.

On the other hand, this mitigation strategy presents risks and challenges in implementation. When reallocating inventory, either within theater or from another CCMD, there is a risk of a depleted supply at the donating location, making that location vulnerable to meeting its conflict demands should they arise. Since this relies on taking supply from other areas, the combatant commander of the donating location may resist reallocating that location's supply to meet unexpected demand in another AOR. Additionally, it is important to note that this is a short-term solution that may not actually meet the demands of the conflict scenario; it also relies entirely on sufficient stock inventoried across locations. If available munition stocks are insufficient to meet the demands of the conflict, this mitigation strategy will only address a fraction of the required demand.

This strategy heavily relies on transportation capacity, which may not be guaranteed in a conflict scenario. In Operation Desert Storm (1991), RAND research indicates that while munition supply was sufficient to meet the unexpected demands of the conflict, the USAF heavily relied on Saudi Arabia's infrastructure to move munitions quickly.[3] As logistics and transportation are a central element to this mitigation strategy, insufficient infrastructure may be a significant barrier to the success of this strategy. In addition, the United States is adherent to local rules and regulations or memorandums of understanding regarding the movement of materiel within a foreign nation. It may be the case that movement of munitions and ordnance is permitted only through certain ports and routes in order to minimize risk to critical civilian infrastructure like roads, bridges, tunnels, and ports. There may also be limitations in the quantity and throughput of explosive materials permitted at any given time. These factors limit the selection of ports of embarkation and debarkation and the ability to move munitions within another country. These regulations could also lengthen the time needed to transport munitions from one area to another.

Temporarily Expand Industrial Capacity

As demonstrated by several case studies, expanding current industrial capacity to produce munitions may be able to meet a surge in demand in the short term. Examples would be constructing facilities and expanding the workforce as demand arises. Unlike reallocating inventory, this mitigation strategy allows the supply to expand without placing risks in other locations.

---

[3] Pyles and Shulman, 1995.

Some of the key costs of this mitigation strategy include the cost of acquiring new facilities, equipment, or tooling to build munitions in response to a surge in demand. Additionally, this capacity would be intended to meet surge demands, so it would most likely remain idle or produce at low levels afterward, during normal demand, which would require either retaining staff to meet that surge capacity or quickly hiring when a surge in demand arises. This mitigation strategy would require expansion across a significant portion of the supply chain, including upstream suppliers. If capacity is expanded at the final stages of the supply chain, such as integration and final assembly of munitions, this downstream component is highly dependent on other stakeholders in the supply chain. If the capacity of one component of the supply chain affects the ability of other portions of the supply chain to surge and meet demand, this mitigation strategy may not be able to meet the surge demands without additional investment.

Additionally, a strategy that requires hiring personnel and expanding facilities may also require certification and clearances in advance. If personnel are hired as surge demand arises, the supply chain may encounter delays as those individuals need to pass clearances and meet hiring requirements.[4]

Add Production Shifts

Adding production shifts is a mitigation strategy that uses generally the same production capacity of facilities but relies on increasing the labor to expand production capacity. As surge demand arises, the supply chain may hire more individuals to operate production lines in addition to the current labor, perhaps simultaneously or else at hours that are typically not part of normal production.

This mitigation strategy provides a relatively rapid response to potential surge demands but relies on a confluence of factors to meet the demand as it arises. Hiring individuals may require an extensive onboarding process, such as attaining security clearances, and specialized training that may delay their start. Additionally, this mitigation strategy relies on enough slack in the labor market that suppliers can quickly hire staff with the appropriate experience, training, or ability to acquire a security clearance. Depending on the hiring requirements, it may be difficult to find the appropriate individuals to meet production demand whether in a plentiful labor market or during a labor shortage. Like expanding production capacity, this mitigation strategy would most likely need to be applied across a large swath of the supply chain, including upstream suppliers, as certain aspects of the supply chain may be heavily reliant on touch labor or other factors. If one part of the supply chain does not have the capacity to meet the demands of the other, then delays and limits to total production will remain.

---

[4] For additional information on delays generated by hiring and certification demands, see Obaid Younassi, Kevin Brancato, John C. Graser, Thomas Light, Rena Rudavsky, and Jerry M. Sollinger, *Ending F-22A Production: Costs and Industrial Base Implications of Alternative Options*, Santa Monica, Calif.: RAND Corporation, MG-797-AF, 2010.

Finally, this mitigation strategy is reliant on the assumption that material inputs are available and sufficient—that is, production materials need to be quickly acquired for the installed capacity to meet the production demands, given sufficient personnel levels. In some cases, there are competing demands for specific components that are shared across different munitions that place constraints on any specific munition being able to meet its production demands without simultaneously reducing another supply chain's access to that same component. If there is competition for components and materials required to produce the munition, then this mitigation strategy, as well as others, may have a limited ability to meet the surge demand.

Fast-Track Munitions Close to Initial Operating Capability

As demonstrated by the acceleration of JDAM employment in Kosovo, it is possible to quicken the evaluation cycles for new munitions or at least kits. However, it must be cautioned that this was only because much testing had already been done before their employment in Operation Allied Force. Over 450 guided test launches were conducted, and in fact, 937 JDAMs had already been handed over to the USAF the year prior during the delivery of Lot 1.[5] As mentioned previously, the JDAM was still in the testing phase—it did not reach IOC until February 2001.[6]

The applicability of this mitigation strategy depends on the amount of testing done prior to the conflict and the corresponding surge in demand. It may also depend on the degree of technical difficulty in achieving IOC. One may argue that the effort required to reach the appropriate technology readiness level (TRL) for a modification of an existing munition is simpler than the effort required for a completely new munition. The risk may be lower as well. If a munition is a derivative of an existing system, or an upgraded version, it may represent a similar TRL deficit to overcome and risk imposed by that of the JDAM. Additional research is needed to manage the risk associated with accelerating testing and certification of munitions very near IOC.

Mid-Term Strategies (2–7 Years)

Mid-term strategies are less constrained than short-term mitigation strategies in that they allow more time to buy down risk. While they may not solve current readiness issues, mid-term strategies promote modification of the supply chain in a more permanent way when compared to short-term strategies that are likely to be temporary. Within the mid-term time frame, we assume that new munitions cannot be introduced into the supply chain, but there is flexibility to build more and increase overall capacity. In this time frame, force development and planning can also

---

[5] Wolfe, 1999.

[6] U.S. Department of Defense, *Selected Acquisition Report—Joint Direct Attack Munition (JDAM)*, December 2018.

shift, which may affect the supply chain's ability to meet the demands of the changing force. In this time frame, mitigation strategies include increasing inventory and building new factory capacity.

Increase Inventory

Increasing inventory provides opportunities to meet the expectations of operational planning and ultimately address anticipated demands before the outset of a potential conflict scenario. This strategy includes performing acquisitions before surge demand.

The timeline of increasing inventory can vary depending on anticipated surge demand and the production capacity of the supply chain. The acquisition can be planned over the span of several years, which may alleviate any significant production pressures on the supply chain to meet acquisition requirements but may run the risk of not meeting unanticipated surge demand requirements should they arise in the middle of the acquisition. If the acquisition is done up front with a relatively short time frame, it will place pressures on the supply chain to meet the acquisition and inventory requirements. This mitigation strategy also relies on the assumption that the current supply chain has capacity to meet increased production and inventory requirements or that it is able to expand, which may incur additional costs. In either case, this acquisition will require a large up-front cost.

Barriers to implementing this strategy are funding and capacity related. The first barrier is the funding required to conduct the acquisition, as increasing inventory will be costly and is a guaranteed cost. Surge demand may be difficult to predict and, in most cases, is never guaranteed to arise, so increasing inventory increases preparedness to respond to a surge demand but "locks in" investments and funding to a specific munition inventory rather than leaving that for flexible responses in the face of different surge demands that may come to pass.

The second barrier to implementation is a consequence of providing funding to increase the inventory of specific munitions. By locking in those investments for one type of munition or a set of munitions, that funding will most likely be taken away from another program or type of munition. In line with the first barrier discussed, this will increase preparedness to respond to a *limited set* of surge demands that may arise in the future but may lock out or place limitations on the ability to respond to surge demands for other munitions or demands across other capabilities. By increasing inventory and committing to increased production, this limits future flexibility.

Additionally, capacity remains a constraint, given the numerous supply chain challenges identified in the industrial base.

Build New Factory Capacity

Building new factory capacity can expand the production lines available to produce munitions in response to a sudden surge in demand. This strategy requires constructing new facilities, acquiring additional equipment and tooling, and expanding staffing levels. These facilities and additional capacity can either be used in addition to current capacity to increase inventory (as discussed earlier in this chapter), or these facilities may be kept below maximum capacity during peacetime and then ramped up to meet surge demand, although they would still be subject to some of the same labor constraints mentioned earlier. This mitigation strategy may assist the implementation of other mitigation strategies—for example, by providing additional production capacity that could help increase inventory before surge demand.

One of this mitigation strategy's greatest benefits, as well as its greatest weaknesses, is the perceived "permanence" of the investment. Building new facilities requires a large up-front capital expenditure. Should facilities run below maximum capacity, these facilities could be operating at inefficient levels by potentially mothballing equipment and tooling and retaining minimal staff levels, but costs would increase over the long run. When a surge demand arises, there will also be a surge in costs to increase the number of staff, unpack mothballed tooling, and acquire materiel to ramp up production. On the other hand, if the facilities are running at an operational capacity when not in surge demand, the facilities, staff, and tooling will be an additional cost.

Additionally, depending on the facilities and the nature of the munitions produced, this investment may limit the flexibility to respond to surge demand requirements for different munitions. If the facilities and tooling required are unique to a specific munition, then they may be limited in their ability to meet surge demands that may not be appropriately addressed through that munition. This strategy may better address a wider range of surge demand scenarios if the facilities, equipment and tooling, and staff are able to address surge demands that are applicable to a broader range of munitions or other capabilities. Thus, this mitigation strategy should be implemented by expanding a production line that manufactures a common part of multiple munitions or a production line capable of manufacturing multiple munitions or munition components. The shared production line for the LRASM and JASSM-ER is one example.

Long-Term Strategies (5–15 Years)

Long-term mitigation strategies are recommendations that change the overall nature of the munitions and, consequently, the supply chain. With a long-term time frame, there is flexibility in shaping the force design and program of record. These mitigation strategies include introducing and expanding modular designs across the supply chain and incorporating additive manufacturing. Because of the length of the acquisition process, particularly the technology maturation and risk reduction phase and the engineering and manufacturing development phase, the "time to impact" of these strategies is longer than most.

Modular Designs

Modular designs are greater systems, such as a munition series, broken into smaller, independent systems (modules) that can be linked and exchanged in various combinations to produce different munition products using a shared platform. With modular systems, a broader number of common parts across different systems could potentially be reallocated to address surge demands quickly—assuming the parts are interchangeable across different systems and in stock. Modular designs allow individual components within a platform to be swapped and maintained independent of the other components of the broader system. Modular designs also allow for easier upgrades.

Using modular designs to maintain and upgrade systems may provide opportunities to reallocate personnel typically focused on sustainment to production—thereby increasing the number of available personnel to address surge demands. Additionally, a modular system may generate long-term cost savings, since having a common design may reduce development costs. Rather than redesigning specialized components for each new system, previously designed components are leveraged. Another area for cost savings may come from increased market competition at the subcomponent level. Instead of bidding on an entire platform, would-be suppliers could compete to provide modules to the shared platform. This may also reduce risks of cultivating single-source suppliers within the munition supply chain.

Modular designs have been used in the past to help mitigation strategies, albeit to a relatively minor extent. In Operation Inherent Resolve, JDAM modularity was touted as an opportunity to quickly assemble munitions in the field.[7] With modular systems in munitions, stakeholders are able to make rapid changes to the munitions—such as adding or removing a laser sensor kit on a JDAM—in the field to match the demands of the conflict or mission.

It is also important to note that modular concepts are in development across the DoD. The Air Force currently has a Modular Advanced Missile program, though the degree of interchangeable components is currently not available in public resources.[8] Modularity is not a new and innovative mitigation strategy, but there are further opportunities to develop the nature and degree of modularity in systems to enable a strategic capability to rapidly respond to surge demand.

Developing modular designs will likely create up-front costs in both developing the common platform and the modules to meet the requirements of the USN or other services. The Joint Capabilities Integration and Development System process is likely to be lengthy in making sure the right requirements are in place. Depending on the degree of modularity desired and the ultimate deviation from current systems, the costs of the overhaul could also include training new, specialized production staff or retraining current staff. If the modular systems leveraged existing production lines, they would also have to accommodate old designs along with adapting old equipment or acquiring new equipment.

---

[7] "Boeing Ramps Up Bomb Production as Stockpiles Decrease," *Military.com*, March 29, 2017.

[8] Steve Trimble, "The Weekly Debrief: More Details Emerge About New USAF Mystery Missile," *Aviation Week & Space Technology*, April 5, 2022.

The challenges of introducing new capabilities into the supply chain are a barrier to implementation. It may be difficult to acquire the substantial funding that may be required for a long-term investment to transition to modular systems. Additionally, OEMs may be hesitant to invest capital to develop and produce new capabilities. Modular systems and the overall degree of modularity desired may dramatically shift commonplace munition supply chain practices and change the face of the supply chain—and it may be a slow process. Adding to the overall time is the defense acquisition process itself. To properly develop a shared platform from which to add subcomponents, the shared platform would have to be designed in such a way that it met the requirements for a variety of mission sets. To be truly modular, the platform would need to perform in multiple use cases—long range and short, heavy payload or light, air intercept or air to ground, and so on. The platform would need to be certified in each of its configurations whenever a new modular component is proposed. It is therefore unlikely that a modular munition would reduce the time needed for certification by the testing and evaluation community.

Additive Manufacturing

Additive manufacturing (AM), otherwise known as "3D printing," may create opportunities for different production methods across the munitions supply chain. This method of manufacturing works through layers of "printed" materials that allow more precision in the design of individual components as well as the overall system and synthesis of different materials throughout the printing process. A benefit of this process is the ability to rapidly develop and modify parts in the supply chain, as the ability to print layer-by-layer presents an opportunity to create structures and designs not feasible with machining and other subtractive methods.[9] With this ability to produce new geometric structures, it is possible to reduce the total number of components. A component that once required complex welds or consumables, like nuts and bolts, can be consolidated into a single part. This reduces weight, part count, and materiel costs.[10] Reductions in weight can also translate into fuel savings and increases in range in the right platforms. Additionally, the printers used in AM can quickly switch between product lines, creating a flexible production capacity to meet demands as they arise. The nature of AM and the layered production approach allows the same equipment to be used to manufacture different components. There are also fewer changeover requirements to prepare the manufacturing equipment to produce new parts with different materials. This also affects the ability to rapidly respond to demand for spares and parts. Instead of having to retool an entire production line, a few printers could be quickly reconfigured with the proper materials and component blueprint to produce the requisite quantity of spares.

---

[9] Whitney Hipolite, "3D Printed Guided Missiles are Now a Reality Thanks to Raytheon," *3DPrint.com*, July 16, 2015.

[10] S. W. Williams, F. Martina, A. C. Addison, J. Ding, G. Pardal, and P. Colgrove, "Wire + Arc Additive Manufacturing," *Materials Science and Technology*, Vol. 23, No. 7, February 9, 2016.

There are a few contemporary efforts aimed toward implementation of AM for aerospace and defense with applicability to munitions. The Air Force Research Laboratory's Eternal Quiver program aims to produce explosive compounds in solid rocket motors using AM.[11] Should this effort come to fruition, explosive compounds could be printed into ordnance and propellant grains for munitions. Other materials, such as metal alloys, polymers, ceramics, and even composites, also lend themselves to AM. Additively manufactured metal components in turbine engines have already been demonstrated; take the LEAP (Leading Edge Aviation Propulsion) engine from CFM International as an example—it uses fuel nozzles produced by metal 3D printing to replace the previous component that was made from 20 individual parts.[12]

This production system allows an overall simplification to the manufacturing process and the supply chain. Standardizing the production of different components to the AM process may also reduce the need for specialized certifications and qualifications that may be required to produce different munitions components today.

The primary barriers to implementation of this mitigation strategy are funding, the magnitude of change, limitations in application of the technology, and time. With such a large change to the nature of production, there will most likely be a significant up-front cost. These costs could include the expenses of redesigning the production system around AM capabilities and requirements; acquiring the facilities, systems, equipment, and personnel; and conducting research and development for the overall process as well as individual component design. In line with the dramatic change this will create across the supply chain, it may be difficult to garner support across the supply chain, as it will be a fundamental shift away from current practices and may carry risks. The risks of applying AM within specific supply chains and munitions production may differ as well, as there will most likely be limitations to where and how AM can be successfully applied in the supply chain. There may be current processes or unique systems that cannot be replicated by the AM process. For example, current AM faces challenges in connecting all the additively manufactured components together while being made of a variety of different materials.[13] It may be that munition integration and assembly will always require some touch labor somewhere in the supply chain. Finally, it is important to note that the process of convincing OEMs to incorporate AM will be slow, and if one printer is responsible for producing different components, there may be barriers to meeting simultaneous demand for a diversity of components or systems. To avoid this problem, more printers would need to be acquired and more money spent.

---

[11] AFRL/RQRM, "Eternal Quiver Industry Day Primer," presentation, July 2020.

[12] Timothy M. Persons, "3D Printing: Opportunities, Challenges, and Policy Implications of Additive Manufacturing," *GAO-15-505SP-Addictive Manufacturing Forum*, June 2015.

[13] Hipolite, 2015.

Summary

Munitions differ from most other supply commodities in that they are wartime critical, with limited use in peacetime. While demand is known, stockpiling and investing in a "use only in case of war" commodity is a monetarily costly decision. However, as has been identified, the current inventory in munitions is critically low for anticipated demand in a conflict against a near-peer adversary. Mitigation strategies for overcoming munitions supply chain challenges should be analyzed across three time horizons (near-term, mid-term, long-term) in conjunction with a risk assessment of the likelihood of conflict with a near-peer competitor. Were conflict to break out in the short term, there is little time for investment in additional inventory or emerging technologies. Therefore, short-term mitigation strategies focus on reallocation of munitions and surge of production. In the mid-term, there is greater ability to invest in more inventory. And in the long term, it is recommended that the Navy invest in emerging technologies.

4. Current State of the Supply Chain for Naval Aviation Repair Parts

A review of the Navy's Class IX[1] naval aviation supply chain highlights that the system is not currently postured to meet demands expected under DMO, particularly when it comes to requirements determination and procurement. The Navy has worked to address issues within the supply chain, but these initiatives focus on fixing near-term readiness and do not account for long-term readiness issues that become more acute under DMO conditions. Current models and demand estimates do not accurately account for DMO requirements, and misaligned incentives among key stakeholders—such as the DLA, the Navy, and industry—make it challenging to source adequately to meet wartime demands. Additionally, the Navy is not currently buying to meet demand required in DMO, but even if it could fund these needs, issues in the industrial base capacity would prevent its ability to surge to meet demand. Further complicating the ability to buy to wartime needs are complex funding mechanisms and reliance on WCF for funding. This chapter focuses on challenges in the Class IX supply chain, and Chapter 5 will outline strategies to mitigate those challenges.

We focused on aviation spares because we believed that they might show the greatest variability between peacetime and wartime use. However, we were hampered by the simple fact that data for wartime use is not known, making comparison primarily a matter of inference rather than a matter of data comparison. In subsequent chapters, we will describe possible ways of improving data collection. For purposes of this report, we have little direct data to rely on; however, we do know from previous conflicts that the types of parts aviation units need do vary between peacetime and wartime use.[2]

Navy Approach to Class IX

The Department of Defense defines Class IX as "repair parts and components to include kits, assemblies, and subassemblies (repairable and nonrepairable) required for maintenance support of all equipment."[3] Class IX management includes requirements determination (for steady-state and wartime demand), procurement, repair, storage, and transportation. For the purpose of our report, we focused on the front end of the management process, which encompasses determining requirements and procuring parts. The Navy determines its spares and repair parts allowances

---

[1] Class IX refers to the military class of supply that consists of repair parts required for maintenance of systems.

[2] Pyles and Shulman, 1995.

[3] Joint Chiefs of Staff, "Joint Logistics," Joint Publication 4-0, updated May 8, 2019.

using the readiness-based sparing (RBS) model as outlined in OPNAV Instruction 4442.5A, and demand-based methods are used when there is not enough data to use RBS methods. RBS methodology determines the most cost-effective allowances to ensure readiness objectives (A_O)[4] and thresholds using analysis and empirical maintenance data.[5] An RBS analysis is conducted at the start of a new acquisition program and is intended to be updated throughout the life cycle of the program through annual assessments. Key metrics that determine the system's performance include achievement of A_O and logistics performance as measured by customer wait time. This emphasis on cost-effectiveness and readiness that is based primarily on historical data drives the system to prioritize procuring and stocking parts that address current or historical maintenance issues rather than anticipating and balancing parts needed for future demands.

The naval aviation community has experienced challenges in recent years with a decrease in annual mission-capable rates for Navy aircraft. A review of issues affecting these low rates highlighted critical supply and maintenance challenges, such as unexpected replacement of repair parts due to aging aircraft, service life extension, depot delays, shortage of trained maintenance personnel, diminishing manufacturing source, parts obsolescence, and parts shortages and delays.[6] The Navy has launched initiatives to address critical performance and supply chain challenges in the fleet. They include but are not limited to Performance to Plan (P2P), Naval Sustainment System–Supply (NSS-S), Integrated Supply Chain Management (ISCM) Control Tower, and Naval Performance Improvement Educational Resource. These initiatives are collaborative in nature, attempting to integrate stakeholders across the supply chain and helping to eliminate historical stovepipes within the system. The emphasis is on data-driven approaches to achieve measurable results.

These efforts have proved effective at regaining near-term readiness for naval aviation. For example, the ISCM Control Tower has increased ready-for-issue inventory for F/A-18s at flight line by 21 percent and reduced the number of aircraft awaiting critical components by 8 percent.[7] Under NSS-S, the Naval Supply Systems Command (NAVSUP) Weapons Systems Support (WSS) Integrated Weapons Support Team was tasked to address unfilled portfolio orders for F/A-18 Super Hornet generator converter units. In their review, the team found that production at the OEM was dealing with competing demands between Naval Air Systems Command, Boeing,

---

[4] A_O is defined as "a measure of the degree to which an item is in an operable state and can be committed at the start of a mission when the mission is called for at an unknown (random) point in time" (Department of the Army, "Army Regulation 702-19: Reliability, Availability, and Maintainability," February 12, 2020.)

[5] Department of the Navy, "NAVSUP Weapons Systems Support Instruction 4441.15L," June 26, 2017.

[6] Government Accountability Office (GAO), "Weapon System Sustainment: Aircraft Mission Capable Rates Generally Did Not Meet Goals and Cost of Sustaining Selected Weapon Systems Varied Widely," GAO-21-101SP, November 19, 2020.

[7] Brian Jones, "Integrated Supply-Chain Management Pays Off for Naval Supply Systems Command, Partners," U.S. Navy Press Office, November 5, 2020.

and WSS.[8] The team managed these demands by once again baselining the contract schedules, which resulted in WSS achieving zero unfilled customer orders in the portfolio. Re-baselining contract schedules assists in increasing communication among agencies competing for parts but does not address the challenge of the manufacturer being unable to meet the contract demand of all stakeholders. Under more demanding operating scenarios, schedules most likely will not allow the flexibility that steady-state scheduling supports. NSS-S may need to look at other mitigation strategies to increase order support if the expected demand for generator converter units will be robust under wartime conditions.

The application of commercial best practices allows the Navy to set baselines for internal management of supply chains, as well as set expectations for other stakeholders in the Navy supply chains. While this allows the Navy to improve its supply chain velocity and establish proven performance standards, it does not explicitly address potential surge demand in a conflict scenario.

Conflicting Incentives Among Stakeholders Challenge Supply Chain Support

The supply chains for naval aviation repair parts contain a multitude of stakeholders with differing incentives. These incentives are nonaligned, making it difficult to support future wartime demands because there is a lack of incentives to provide surge capacity. The primary stakeholders external to the Navy supporting naval aviation parts requirements determination and parts procurement include DLA and the industrial base.

DLA serves as the combat logistics support agency for the military services, combatant commands, and when required, other government agencies as well as allies and partner nations. DLA is responsible for managing the "end-to-end global defense supply chain—from raw materials to end user disposition."[9] To fulfill its mission, DLA is strongly incentivized to prioritize efficiency at a reduced cost balanced with responsiveness to warfighter demands.[10] DLA measures its performance primarily on how well it maximizes materiel availability (MA), which demonstrates its ability to meet customer demand. For some commodities such as food and medical supplies, MA is high (e.g., 99 percent), since these commodities have predictable demand, can be ordered in high volume, and are commercially available because they are not military unique.[11] Repair parts, however, are less predictable, smaller volume, and oftentimes military unique.

---

[8] Brian Jones, "NAVSUP WSS Integrated Weapons Support Team Slashes Unfilled GCU Orders to Zero," *DVIDS*, October 29, 2021.

[9] Defense Logistics Agency, "About the Defense Logistics Agency," 2022.

[10] Paul H. Richanbach, H. T. Johnson, Kathleen M. Conley, Graeme R. Douglas, Michael F. Fitzsimmons, Kenneth A. Evans, and David R. Graham, *Independent Review of the Defense Logistic Agency's Roles and Missions*, Institute for Defense Analyses, December 2014.

[11] Richanbach et. al, 2014.

Therefore, to achieve high MA for repair parts, the DLA system prioritizes stocking less expensive parts with predictable demands with minimal excess inventory. Expensive parts with infrequent demands are allowed lower MA levels. While MA is a useful metric from a business perspective, a focus on improving MA does not necessarily show how well customer need is met, particularly for critical parts.[12] Additionally, it focuses the organization on meeting steady-state requirements rather than being optimized to meet future demand.

Recognizing this, starting in 2018, DLA operations directorate has stood up the service demand readiness summit to capture anticipated demand for Class IX (maritime, aviation) for future operations.[13] The summit provides an opportunity for the services to convey their priority and demand forecasts to DLA. Without this service-level input, the system defaults to historical input. The summit can inform a manual change in the operating system to dial up or down forecast levels balanced with fiscal constraints, but it is not integrated into formal forecasting models.[14] While this effort does look at anticipated demand, discussions with DLA highlighted that forecasting discussions are still focused on the near term. What is missing is identification of requirements to support operations plans (OPLANs) and stock against those.

The industrial base is optimized to maximize profits and revenue for their businesses. Business decisions therefore are aligned toward these goals. For some systems, such as the F-35, defense contractors are responsible for some portion of repair parts management through performance-based logistics contracts. These contracts are written to incentivize contractors to achieve a certain level of system availability rather than on the number of transactions. Contractors are incentivized to increase reliability, and financial gain is achieved when contractors use a smaller number of spares and repair parts. However, the organizational infrastructure and culture of most contractors are optimized for transactional, profit-maximizing execution—particularly if the military portion of the business portfolio is a small fraction of the overall business, which is the case for many of the defense and aerospace contractors.[15]

This shift in placing some portion of spares management to the contractors is most easily seen in the case of the F-35. The F-35 uses a complex spares program that involves the use of pooled spares for global and base spares, as well as deployment and afloat-based spares packages. The prime contractor is responsible for managing the spares supply chain and is held accountable to availability metrics. However, a recent review of the F-35 supply chain has

---

[12] Marc Robbins, James R. Broyles, Josh Girardini, Kristin Van Abel, and Patricia Boren, *Improving DoD's Weapon System Support Program*, Santa Monica, Calif.: RAND Corporation, RR-2496-OSD, 2019.

[13] Discussion with Defense Logistics Agency officials, September 2021

[14] Discussion with Defense Logistics Agency officials, September 2021.

[15] James Marceau, "Viewpoint: Performance Based Logistics Contracting—Does It Work?" *National Defense*, August 8, 2018.

identified numerous challenges with the F-35 meeting its performance requirements because of spares shortages and repair backlogs.[16]

Distributed Maritime Operations Will Create Different Demand Profiles for Aviation Repair Parts

The current process of sparing takes a historical approach, but recent requisition history is likely much lower than the quantities needed in a demanding wartime scenario. Determining the difference between steady and surge demand requires an analysis of OPLANs and defense scenarios. Getting to actual demand under these scenarios is difficult, but case studies and interviews with subject-matter experts provide indications of how demand may change from steady state to wartime conditions using DMO, particularly in a demanding operational scenario, as envisioned in the Western Pacific or the European theater.

DMO is the Navy's concept to meet the demands expected in a return to great power competition, as outlined in the National Security Strategy and National Defense Strategy, where China and Russia are the primary pacing threats. The Navy's concept relies on distribution of naval assets where every platform is a sensor or shooter contributing to the fight. This distribution of forces across the battlefield has implications for naval aviation assets that will likely fly longer sorties and will see a sharp and constant demand for mission-capable aircraft.

With longer sorties (increasing up to ten hours), flight hours will accumulate more rapidly and thus phased maintenance requirements occur more often. The increased demand for mission-capable aircraft will make cannibalization of frontline aircraft undesirable.[17] Increases in aviation units called forward will also mean the Navy may not be able to rely on cannibalization of "home station" aircraft as it has done in previous conflicts. Fighting a near-peer competitor will likely incur battle damage not seen historically—requiring swap-outs of large portions of the aircraft, such as wing assemblies.[18] Additionally, the aircraft will be flown differently than they are in peacetime. For instance, electronic systems are often not turned on during peacetime operations and will receive significantly more use under combat operations. Previous research on surface ships indicates that systems rarely used in peacetime fail in unexpected ways when subjected to conditions similar to wartime.[19] These factors will combine to create a different demand profile

---

[16] Government Accountability Office, "F-35 Aircraft Sustainment: DOD Needs to Address Substantial Supply Chain Challenges," GAO-19-321, April 2019.

[17] Cannibalization refers to the removal of serviceable parts from one weapon system to replace unserviceable parts on another system. It is a method usually used when the expected delivery date for the replacement item is delayed due to supply issues and waiting for the item may incur operational risk.

[18] Discussion with retired aircraft maintainer, October 2021.

[19] Bradley Martin, Roland J. Yardley, Phillip Pardue, Brynn Tannehill, Emma Westerman, and Jessica Duke, *An Approach to Life-Cycle Management of Shipboard Equipment*, Santa Monica, Calif.: RAND Corporation, RR-2510, 2018.

for naval aviation repair parts than what has been seen historically and is feeding current sparing models.

Air operations during Desert Storm provide a useful case study of aircraft break rates and logistics implications under non-peacetime operations. During Desert Storm, the Air Force experienced a twofold increase in break rates across its aircraft platforms due to more stressful sorties (higher-than-planned sortie rate, longer sorties), which induced greater stress on some subsystems. Additionally, some mission-critical subsystems were exercised more. For example, F-117A pilots used their inertial navigational system more for longer sorties, but F-111Es and Fs did not require their terrain-following subsystems because they flew more mid- to high-altitude missions.[20]

Likewise, during Desert Storm, break rates were not consistent across aircraft platforms— some break rates increased, some decreased, and others remained constant. Break rates by aircraft also varied across the duration of operations. For instance, F-15Cs saw immediate increases in their break rates on deployment most likely because the mission had to fly combat air patrols (CAPs) immediately. Pilots flying CAPs used all available systems, thus exercising them more than in peacetime.[21] For F-16Cs, mission restrictions were severe in the early phases until bombing ranges and training sorties could be established.[22] Thus, break rates were less than those at home station in the initial deployment period but ramped up significantly during early operations. This indicates that techniques that assume a proportional increase in workload and maintenance demand to operational tempo miss the nuance of changes in break rates. The inherent unpredictability of wartime demands that logistics support cannot rely solely on proportional formulas but rather requires better engineering assessments that account for changes in system usage in DMO conditions.

Limitations of Current Industrial Base Capacity and Stockpiling Prevent the Ability to Surge

To meet the requirements likely under DMO, the Navy will probably need to stockpile repair parts or be able to surge industrial capacity to meet demand. Analysis of naval aviation supply chains indicates that there are vulnerabilities in the industrial base that will limit the Navy's ability to surge. For example, the Department of Defense Inspector General's review of five critical supply parts for the F/A-18 determined that the Navy and DLA could not obtain enough to meet current backlogs and projected demand because of the obsolescence of parts, single

---

[20] Pyles and Shulman, 1995.

[21] Pyles and Shulman, 1995.

[22] Pyles and Shulman, 1995.

vendors, and other industrial base challenges.[23] Other reporting indicates that the F-18E/F is facing 18 known cases of diminishing manufacturing sources and materiel shortages in the next two years.[24] While this reporting provides indications of where the industrial base may be facing supply chain challenges, the reality is that data on suppliers below the prime contractor is fragmentary,[25] and the Navy lacks adequate resources to track these supply chain issues. Furthermore, the industrial base for spares is complicated, and the ability to create "warm lines" is problematic. Determining where to prioritize investment in factory capacity will take more research.

DoD and DLA have some programs in place to mitigate some of these challenges. Of particular note is the DLA's warstopper program. This program allows the government to invest in wartime capabilities where readiness requirements are higher than the industrial base is willing to invest. This capability was started in 1993 based on an after-action item identified during Desert Storm to shore up at-risk manufacturing capability of an auto-injector used to deliver nerve-agent antidote. It is now used annually to fund risk-mitigating investments for critical wartime items and sectors.[26] According to the Industrial Capabilities Report to Congress, the warstopper program has three primary functions:

–Secure commercially available go-to-war material in the quantity and timeliness [needed] (example: pay management fees to guarantee the quantity and early delivery)

–Increase manufacturer and distributor capability to provide go-to-war consumable items material (example: stage raw material and long lead time parts or provide additional equipment)

–Preserve cold production needed for go-to-war consumable items (example: fund a company's fixed cost to sustain a production line).[27]

While the warstopper program provides a vital capability, it is challenged by the fact that warstopper items are not always easy to predict. For aviation, the priority is to invest in raw material buffers such as steel and titanium,[28] which provides more flexibility. Use of the other mechanisms has been less evident, suggesting further research into areas where preserving cold

---

[23] U.S. Department of Defense Inspector General, *Audit of Navy and Defense Logistics Agency Spare Parts for F/A-18 E/F Super Hornets*, DODIG-2020-030, November 19, 2019.

[24] Gregory E. Saunders and Nicole Dumm, "Diminishing Manufacturing Sources and Material Shortages," *Defense Standardization Program Journal*, October/December 2017.

[25] Caolionn O'Connell, Elizabeth Hastings Roer, Rick Eden, Spencer Pfeifer, Yuliya Shokh, Lauren A. Mayer, Jake McKeon, Jared Mondschein, Phillip Carter, Victoria A. Greenfield, and Mark Ashby, *Managing Risk in Globalized Supply Chains*, Santa Monica, Calif.: RAND Corporation, RR-A425-1, 2021.

[26] Dianne Ryder, "Rare but Ready," Defense Logistics Agency, December 26, 2016.

[27] U.S. Department of Defense, *SRM Industrial Capabilities Report to Congress: Redacted Version*, June 2009.

[28] OSD A&S Industrial Policy, *Fiscal Year 2020 Industrial Capabilities Report*, January 2021.

production or increased manufacturing capabilities for system components may be beneficial. The warstopper program is also challenged by the need for close collaboration between suppliers and contractors, as these funding mechanisms are often outside normal contractual practices that vendors are comfortable with.

Similarly, stockpiling provides a mitigation strategy for meeting surge demand; however, the services are reluctant to stockpile absent a clear indication that something needs to be stockpiled, and DLA lacks legal authority and incentives to stockpile absent service funding to do so. DoD pursues two different approaches to stockpiling. One approach is the National Defense Stockpile, which is based on raw materials. The other program is the war reserve materiel (WRM) stocks. These stocks are managed by the services and consist of principal end items, secondary items, and munitions to buffer against requirements for contingencies and scenarios. The Navy previously had a WRM program whose stated purpose was "to provide the additional materiel, over and above peacetime operating and training stocks, needed to support the force structure dictated by the [Secretary of Defense] planning guidance."[29] Materiel was held in supply accounts to serve as "swing stocks" applied to any scenario.[30] However, the Navy discontinued the program in 2011. This indicates that the Navy has prioritized lower-cost options to mitigate risk, as stockpiles can be costly endeavors.

Understanding the cost implications, buffer stocks still are an effective tool for mitigating incorrect forecasting, given the uncertainty of the wartime environment, but a detailed analysis of which type of resources are most suitable to buffer is recommended. Buyout strategies are often not feasible economically, but research has shown the effectiveness of tailored partial-buying strategies for spares when considered across frequency of demand, criticality for mission success, cost, weight, and availability.[31]

Working Capital Fund Arrangements Do Not Effectively Provision Low-Demand but Possibly Critical Supplies

DLA is funded to perform its mission through a WCF construct, which is a rolling fund arrangement that allows DLA to purchase items as measured service demand dictates.[32] The

---

[29] Office of the Chief of Naval Operations (OPNAV), Navy War Reserve Materiel Program, Office of the Chief of Naval Operations Instruction 4080.11D, January 21, 1999.

[30] Kristin F. Lynch, Anthony DeCicco, Bart E. Bennett, John G. Drew, Amanda Kadlec, Vikram Kilambi, Kurt Klein, James A. Leftwich, Miriam E. Marlier, Ronald G. McGarvey, Patrick Mills, Theo Milonopoulos, Robert S. Tripp, and Anna Jean Wirth, *Analysis of Global Management of Air Force War Reserve Materiel to Support Operations in Contested and Degraded Environments*, Santa Monica, Calif.: RAND Corporation, RR-3081, 2021.

[31] Irv Cohen, John Abel, and Thomas Lippiatt, *Coupling Logistics to Operations to Meet Uncertainty and the Threat*, Santa Monica, Calif.: RAND Corporation, R-3979-AF, 1991.

[32] U.S. Code, Title 10, Armed Forces § 2208, Working Capital Funds, 1994.

intent is to balance out costs and revenue over the budget cycle. This arrangement has many advantages when assessed from a purely business case. The WCF also is set up so that the services are paying only for the things known to be needed. DLA can purchase large numbers of like supplies from companies that have an incentive to maintain capacity for a well-understood level of demand. This works for meeting Navy requirements for day-to-day needs and continuing operations.[33] WCF spare levels are generally effective in providing the required level of support: Replacement for things that break frequently are in ready supply and are purchased in economic order quantities. However, current planning for the Navy is for a condition where demand may be considerably higher than it would be in normal operating conditions.

The WCF aims for a net zero balance on a fiscal year basis, which puts limits on the ability to plan for long-term wartime demand. Currently, the Navy's WCF is in fluctuation, with negative balances projected into the future. At the end of FY 2021, the fund had a cash balance of negative $1.1 billion.[34] To rebalance the fund, the Navy has pursued multiple efforts, from redesigning the pricing system to introducing outfitting assurance and standing up the NAVSUP WSS cash war room. The focus of these efforts has been on reviewing parts with the intent to delay, defer, reduce, and validate the requirement for current fiscal year supply management WCF savings.[35] All of this has the potential to further defer funding for parts required to meet wartime and surge demand.

Research on other WCFs offer insight and potential means to increase readiness for wartime demands.[36] Wartime demand differs from everyday demand, creating a friction with funding readiness of wartime-specific functions and peacetime demands in a WCF. Rates can reflect the activity's effect on readiness: Services essential to readiness should be priced competitively with commercial rates. Subsidization of rates and price transparency can help incentivize customers to support the WCF rather than seek out alternative suppliers.

---

[33] Discussion with Defense Logistics Agency officials, September 2021.

[34] CAPT Scott Stahl, "Naval Sustainment System—Supply (NSS-S) Navy Working Capital Fund (NWCF) Optimization Pillar," *The Navy Supply Corps Newsletter*, Summer 2021.

[35] Brian Jones, "Naval Supply Systems Command Cash War Room Keeping NWCF Solvent," *U.S. Navy Press Office*, September 17, 2021.

[36] The focus of these reports is on the Transportation Working Capital Fund (DLA) and WCFs relevant to the Defense Finance and Accounting Service. See Kathryn Connor, Michael Vasseur, and Laura H. Baldwin, *Aligning Incentives in the Transportation Working Capital Fund: Cost Recovery While Retaining Readiness in Military Transportation*, Santa Monica, Calif.: RAND Corporation, RR-2438, 2019; Edward G. Keating, Ellen M. Pint, Christina Panis, Michael H. Powell, and Sarah H. Bana, *Defense Working Capital Fund Pricing in the Defense Finance and Accounting Service*, Santa Monica, Calif.; RAND Corporation, RR-866, 2015.

Summary

Analysis of existing naval aviation Class IX supply chains indicates that there is a bias toward recovering near-term readiness at the expense of investment in long-term wartime surge capabilities. This is exhibited in the methods the Navy uses to forecast repair parts, the incentive structure for key stakeholders, and the funding mechanisms the Navy employs to buy repair parts. Estimates of what is required to support DMO in a large-scale conflict against a near-peer competitor indicate that what the Navy is currently buying is not what the Navy will need to support operations. To address these challenges, mitigation strategies must overcome near-term bias and be tailored in their approach.

5. Mitigation Strategies for the Naval Aviation Repair Parts Supply Chain

"Class IX" covers a wide variety of components and products, from very complicated assemblies to simple consumables. To a degree, the demand for spare parts is relatively easy to establish in normal operating environments. Parts are replaced as they wear out, and the demand is tracked through the WCF mechanism. The annual estimated budget for parts is based on a history of annually observed demand.

However, as we will describe in this chapter, this model only works in steady state. There are good reasons to suspect that the demand for parts will change during periods of higher operational intensity, and there are additional reasons for suspecting that the existing store of parts will be insufficient. Addressing these issues in a "to be" framework will take procedural changes and investment. We will examine these issues in more detail, looking first at methods for improving demand prediction and then at the mechanisms for funding parts purchases. Both need to change if we hope for adequate provision of spares in the wartime environment.

The Navy Recognizes Weaknesses in the As-Is Model and Has Taken Action to Correct Them

The Navy is committed to readiness and has made considerable effort to improve parts support across all warfare areas. For example, the P2P and NSS programs are intended to improve predictive maintenance and data collection.[1] P2P is intended to find root causes of current performance, which may include key parts and other leverage points for improved performance. NSS is described as "a follow-on effort that brings in experts from industry or elsewhere to use their own best practices and lessons learned to help the Navy address those specific issues identified in P2P."[2]

Specific to naval aviation, in 2019, the Navy worked to improve fixed-wing tactical aviation mission-capable aircraft numbers from 250–255 to 341 by using the Maintenance Operations Center: Aircraft on Ground (MOC AOG) concept.[3] This is a collective of Navy supply, maintenance, and engineering specialists and industry partners whose activities include identifying the most critical readiness parts for both intermediate and organizational maintenance.

---

[1] Megan Eckstein, "VNCO: Enthusiasm over Navy's Performance-to-Plan Readiness Effort Is Spreading," *UNSI News*, May 24, 2021.

[2] Eckstein, 2021.

[3] Devin S. Randol, "Commander of NAVSEA Visits MOC AOG," Commander, Naval Air Force Atlantic (AIRLANT), March 4, 2021.

The eventual impact of these initiatives is yet to be determined. However, these are best seen as efforts to improve current readiness, leveraging what can be determined from current processes and current demand. Such efforts will no doubt improve predictive capability, but they may not answer the question of dealing with the unknown environment of major-theater war.

Accurately Capturing Wartime Demand Is Critical

Equipment is designed with specific operating parameters in mind and with the recognition that parts will fail during normal use. In general, organizations, including the U.S. Navy, view such replacement as an operating cost. While there are cases where parts are replaced, whether failing or not, as planned maintenance, most parts are simply replaced—and sometimes sent for repair—when they wear out.

The U.S. surface Navy has implemented a "troubled systems" program by which it tries to identify parts and components of parts that are showing higher-than-expected failure rates or having an unusually severe impact on readiness.[4] This program has been led by Naval Surface Warfare Center Corona and is an attempt at a data-based approach to parts management. However, this approach has several limitations: It relies on peacetime data and sometimes just reflects a commander's interest in a particular system. The review might not be systematic and probably will not reveal information about performance in more demanding circumstances.

Moreover, there may be significant differences between surface ships and aviation platforms when it comes to equipment usage in war and peace. Underway surface ships will be operating radars, combat systems, propulsion, and auxiliary equipment. Aircraft might be flying routinely but not activating weapons delivery or combat detection and tracking systems. They will, moreover, likely be operating at a higher operational tempo in combat environments than they would be for routine training. Accordingly, a demand-based approach with steady-state data or even a troubled systems approach relying on current data will generally understate the expected demand.

Improved Engineering Models May Improve Prediction

Equipment installed in ships and aircraft is engineered to a certain level of performance and reliability. With the expansion in ability to process large amounts of data and run large numbers of reliability models, systems commands may be able to better predict failure rates and thus provide a more accurate measure of future demand.

However, even models using big data techniques and with extensive data collection are still subject to limitations from assumptions and data gaps. Predictive maintenance depends on an accumulation of data, and the issue is that the predictions must cover a set of circumstances not

---

[4] Martin et al., 2018.

normally encountered.[5] Our review of the literature indicates that the aerospace maintenance community is looking to use machine learning and other techniques that may identify rare but important failure items. However, the problem here is not that the failures are rare but that the environment is effectively unknown.

A Kill Chain Should Be Used as the Basis for Systems Assessment

The combat models used to evaluate relative systems performance may be of some use because they can identify the systems most essential in kill chains and can thus focus attention on systems most required for combat readiness. An approach that focused on kill chain elements, coupled then with an engineering analysis of similar systems, may have a more regular failure pattern.

For example, when an air-air engagement depends heavily on a search, then a targeting radar, followed by a successful air-air missile firing, the important components of the kill chain are known, and there may in fact be some direct data on elements of the kill chain, such as the search radar. However, there may be little reason to activate the other elements. So, the next effort may be to find components common to those rarely used elements in other systems. Circuit cards, for example, may be used in multiple applications, and information on their performance in other systems may be available.

The next step after that may be to assess the kill chain consequences of a rarely used but still critical element failing. If sufficiently critical but largely unknown and difficult to test, the effort is then either to find an engineering solution or to simply procure spares because the systems are inherently important.

In every case, however, the framing construct is a kill chain, not an observed demand signal. The demand signal must be constructed from the component's place in the supply chain rather than observed demand when the kill chain may not be in use.

Live Testing May Be Essential to Establish Wartime Demand

Live systems testing is expensive and might require the expenditure of large numbers of systems in testing alone. Live testing is frequently done as part of operational test and evaluation prior to a major system being accepted as operationally suitable, but it is not done for tracking and procurement of parts.

To the extent the normal steady-state environment provides a real-world test, additional controlled testing is likely unnecessary. However, as we have discussed, steady state does not describe the real-world environment for rarely used combat systems. Live tests in real-world

---

[5] Maren David Dangut, Zakwan Skaf, and Ian K. Jennions, "An Integrated Machine Learning Model for Aircraft Components Rare Failure Prognostics with Log-Based Dataset," *ISA Transactions*, Vol. 113, July 2021.

conditions approximating combat could provide significantly better engineering and demand models and, in turn, a more realistic assessment of required inventory levels.

Funding Mechanisms Must Account for Investment, Not Just Servicing Current Demand

If steady-state demand were an accurate reflection of wartime requirements, WCF mechanisms would be ideal for fixing spare parts investments. Parts are purchased in accordance with empirically established steady-state demand and may reasonably be expected to be available when needed in sufficient quantity.

Base capacity for spare parts is tied to known demand, and companies have an enduring incentive to make sure they have sufficient capacity. However, as we have determined, steady-state demand is likely to be different from wartime demand, and the inventory levels established by WCF purchases would thus likely be insufficient.

Legislation related to WCFs placed responsibility at the Secretary of Defense level, so the Navy could not on its own decide to cease using WCF as a paying mechanism for the DLA.[6] It may therefore be more practical for the Navy to decide where the construct is beneficial and where the Navy must seek some other approach.[7] In effect, the Navy would use WCF mechanisms for the cases where bulk purchase by the DLA is most economical and effective and use other mechanisms where those are more effective.

Such an Approach Would Require Changes in Program and Budget Submissions. The total demand for spares cannot be thought of as what the Navy is projected to consume in repair parts in a program or budget cycle. What must be procured is the amount required to reliably support kill chains, with demand for materials identified as critical through engineering models, live testing, and recurring instances of mission-incapable supply. This may mean that some parts will be procured wholly in anticipation of a significant but possibly unlikely event, such as a war, and the mechanisms for purchasing these parts will resemble the purchase of weapons.

A possible mechanism could be identifying classes of spares as "kill chain essential" and including their provision as part of program cost and having them purchased directly by program offices on contract from suppliers, without intervening steps from DLA. This might lose some advantages of scale and would force services into supply chain assessments in a way typically not allowed now. However, as we have shown, reliance on DLA and WCF mechanisms are producing situations optimal for the most demanded but generally less critical parts necessary for completing kill chains.

---

[6] U.S. Code, 1994.

[7] G. James Herrera and Brendan W. McGarry, "Defense Primer: Working Capital Funds," Congressional Research Service, updated December 2, 2021.

Sparing for Aviation Systems Should Rebalance Away from Legacy Aircraft to Invest in Future Systems

Legacy aircraft remain a useful part of the U.S. joint force for steady-state operations and will remain so for some period into the future. They are, however, expensive to maintain and generally get more expensive as they age.[8] Moreover, they are generally not useful in expected conditions of near-peer war, so to a degree the question of sparing really is a matter of current versus future readiness.

Accordingly, the path forward for aircraft readiness will include accepting risk in legacy aircraft sparing, with the understanding that this will adversely affect mission-capable rates for these aircraft. Sparing will go to the U.S. Marine Corps and the USN F-35 fleet, with legacy aircraft supported sufficiently for peacetime deployment but not with an assumption of being available for wartime surge.

Related to the shift to prioritizing F-35 wartime requirements for tactical air support, the current contracted sustainment model should be abandoned as oriented toward steady state and replaced with a wartime sparing model without contract incentives. The incentives process for contractor support has all the weaknesses of the DLA model in terms of near-term focus, indeed incentivizing not holding inventory unlikely to be used except in wartime.

Summary

Similar to munitions, mitigation strategies to address supply chain challenges for naval aviation repair parts will need to be considered over multiple time frames. To begin, better engineering data and better demand forecasting through a kill chain system analysis would be beneficial. Addressing funding mechanism issues in the mid-term and rebalancing spares investment away from legacy aircraft and toward next-generation aircraft are also assessed to be of benefit to the service.

---

[8] Matthew C. Dixon, *The Maintenance Costs of Aging Aircraft: Insights from Commercial Aviation*, Santa Monica, Calif.: RAND Corporation, MG-486-AF, 2006.

6. Summary and Conclusions

Our analysis of munitions and spare parts found that the Navy faces numerous challenges and constraints across the supply chain. Figure 6.1. identifies some of those risks, with a particular focus on those challenges affecting the acquisition phase, which is the focus of our report.

Figure 6.1. Challenges Across the Supply Chain Kill Chain

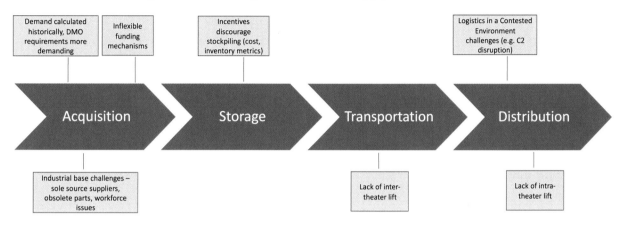

SOURCE: RAND analysis.

Primarily we found that current Navy supply chains are focused on meeting steady-state demands and adjusting to meet near-term readiness concerns. Demand metrics are based on historical analyses that account for decades of conflicts that do not match the expected demands under DMO conditions. Conflict against a near-peer competitor will require different mixes of munitions and spare parts and at likely higher rates than what has been seen historically. While recent initiatives have attempted to better forecast to meet those expected demands, they are not reflected in Navy acquisition and supply systems. In this report, we have estimated demand in the absence of that forecast. Further complicating the matter, differences in incentives among critical stakeholders make it challenging to source to expected demand levels, as estimated by our research team. A shift in focus to just-in-time logistics and resource conservation makes it difficult to stockpile and invest to meet future readiness.

However, even if the Navy were to fix its demand forecasting capability, there are significant issues in industrial base capacity that limit the Navy's ability to surge to meet demand. Diminishing manufacturing sources and material shortages create vulnerabilities within supply chains. Complicated relationships and shared production lines obscure awareness of those vulnerabilities, particularly below the prime contractor level.[1] Funding mechanisms, particularly the use of WCF

---

[1] O'Connell et al., 2021.

to fund Class IX repair parts, make it difficult to buy to future demand or to invest in infrastructure. Therefore, mitigation strategies will need to address planning considerations, budgetary concerns, and industrial base capacity.

Recommendations

Our framework for mitigation strategies assesses options across three time horizons, summarized in Table 6.1. In the near term (0 to 3 years), the focus is on current operations planning and execution. Mitigation strategies include reallocation of inventory from other theaters and surging production shifts to meet any immediate increase in demand for munitions. Likewise, for immediately required surge in repair parts requirements, use of emergency mobilization mechanisms such as the Defense Production Act to incentivize the industrial base could increase production for critical components. For repair parts, mid-term mitigation options (2 to 7 years) should be centered on increasing inventory. Because this time horizon is within the FYDP, many of the recommendations are focused on investment strategies, whether that is buying more munitions, investing in factory infrastructure, or appropriating Navy dollars for buffer stocks. The long-term strategies (5 to 15 years) are focused on investment in future systems and force design. Examples are pursuing modular designs and emerging technology options for munitions, and rebalancing spares investment away from legacy aircraft (e.g., F-18s) and toward next-generation platforms like the F-35.

These mitigation strategies accept risk in the near term in order to invest in increased inventory in the mid-years and rebalance toward newer technologies and system design in the out-years. If a near-peer conflict were to arise in the near term, surge capacity would need to be heavily leveraged to meet expected increase in demand for munitions and spares. As has been demonstrated, relying on surge capacity is only a stopgap. For munitions, reallocation from other theaters will only buy the service time until production capacity can be effectively ramped up. In the case of repair parts, emergency funding mechanisms have proved effective in past conflicts but still take time to implement and for industry to respond.

Table 6.1. Mitigation Strategies by Time Horizon

	Force Employment "Near-Term" (0–3 Years) "Surge"	Force Development "Mid-term" (2–7 Years) "Increase inventory"	Force Design "Long-term" (5–15 Years) "Long-term investment"
Munitions	• Reallocate inventory • Add production shifts	• Build factory capacity • Increase inventory through funding	• Adopt modular designs • Use additive manufacturing
Class IX	• Use emergency mobilization mechanisms (e.g., Defense Production Act)	• Use Navy appropriations to fund "kill chain essential" spares	• Rebalance sparing away from legacy aircraft to next-generation platforms (e.g., F-35)
	• Better calculate demand (better engineering models, kill chain basis for system assessment, live testing)		

Further Research

While this report illuminates supply chain challenges affecting the Navy's ability to conduct DMO and proposes possible mitigation strategies, our analyses were limited by the lack of available data. To effectively implement mitigation strategies, further research is needed to more accurately identify priority areas for investment. We recommend further research, backed by data to better address the challenges identified:

- Conduct a "troubled systems" approach similar to what Naval Surface Warfare Center, Corona has pursued. While we identified limitations of a troubled systems analysis of aviation systems, there still would be benefit in identifying critical systems for further analysis.
- Conduct a "kill chain analysis" for a few critical systems. Such an analysis would be perhaps more useful than a troubled systems approach. Implementing a kill chain analysis for one to two "kill chains" associated with different flight missions would be a way to test the utility of the process.
- Pursue a data-driven review of sparing. Recent analyses done for the Air Force have demonstrated data-driven approaches to partial buyout strategies for sparing that achieve higher readiness to support longer mission durations while being cost-effective. A similar analysis could be conducted for select aircraft systems.

Abbreviations

ALCM	air-launched cruise missile
AM	additive manufacturing
A_o	readiness objective
AOR	area of responsibility
C2	command and control
CALCM	conventional air-launched cruise missile
CCMD	combatant command
DLA	Defense Logistics Agency
DMO	distributed maritime operations
DoD	Department of Defense
FY	fiscal year
FYDP	Future Years Defense Program
GAO	Government Accountability Office
GPS	Global Positioning System
IOC	initial operating capability
ISIS	Islamic State of Iraq and Syria
JASSM	joint air-surface standoff missile
JASSM-ER	joint air-surface standoff missile–extended range
JDAM	joint direct attack munition
LRASM	long-range anti-ship missile
MA	materiel availability
MOC AOG	Maintenance Operation Center: Aircraft on Ground
MST	Maritime Strike Tomahawk
NATO	North Atlantic Treaty Organization
NAV/COMMs	navigation/communications
NAVSUP	Naval Supply Systems Command
NDRI	National Defense Research Institute

NMF	Navy and Marine Forces
NSS-S	Naval Sustainment System–Supply
OEM	original equipment manufacturer
OPLAN	operations plan
OPNAV	Office of the Chief of Naval Operations
P2P	Performance to Plan
PGM	precision-guided missile
PLA	People's Liberation Army
RBS	readiness-based sparing
RF	radio frequency
SRM	solid rocket motor
TACTOM	Tactical Tomahawk
TLAM	Tomahawk land-attack cruise missile
TRL	technology readiness level
USAF	United States Air Force
USN	United States Navy
WCF	working capital fund
WSS	Weapons Systems Support

References

AFRL/RQRM, "Eternal Quiver Industry Day Primer," presentation, July 2020. As of July 11, 2020:
https://beta.sam.gov/api/prod/opps/v3/opportunities/resources/files/736d971838b64672926f7b020007659c/download?api_key=null&token=

"Boeing Ramps Up Bomb Production as Stockpiles Decrease," *Military.com*, March 29, 2017. As of April 25, 2022:
https://www.military.com/dodbuzz/2017/03/29/boeing-jdam-production-stockpiles

CNN, "Inside a Tomahawk Missile Factory," video, YouTube, undated. As of August 17, 2022:
https://youtu.be/FzXW2GZh4qs

Cohen, Irv, John Abel, and Thomas Lippiatt, *Coupling Logistics to Operations to Meet Uncertainty and the Threat*, Santa Monica, Calif.: RAND Corporation, R-3979-AF, 1991.

Connor, Kathryn, Michael Vasseur, and Laura H. Baldwin, *Aligning Incentives in the Transportation Working Capital Fund: Cost Recovery While Retaining Readiness in Military Transportation*, Santa Monica, Calif.: RAND Corporation, RR-2438, 2019.

Cordesman, Anthony H., *The Lessons and Non-Lessons of the Air and Missile Campaign in Kosovo*, Washington, D.C.: Center for Strategic and International Studies, revised August 2000. As of January 10, 2022:
https://csis-website-prod.s3.amazonaws.com/s3fs-public/legacy_files/files/media/csis/pubs/kosovolessons-full.pdf

Dangut, Maren David, Zakwan Skaf, and Ian K. Jennions, "An Integrated Machine Learning Model for Aircraft Components Rare Failure Prognostics with Log-Based Dataset," *ISA Transactions*, Vol. 113, July 2021, pp. 127–139. As of May 12, 2022:
https://www.sciencedirect.com/science/article/pii/S0019057820301750

Davis, C. H., *The JDAM Experience and DPAS*, The Boeing Company, briefing, undated. As of May 13, 2022:
https://www.yumpu.com/en/document/read/40318495/jdam-the-kosovo-experience-and-dpas-dcma

Defense Acquisition University, "Initial Operational Capability," *DAU Glossary*, undated. As of September 9, 2022:
https://www.dau.edu/glossary/Pages/Glossary.aspx#!both|I|27676

Defense Logistics Agency, "About the Defense Logistics Agency," 2022. As of May 12, 2022:
https://www.dla.mil/AboutDLA/

Department of the Army, "Army Regulation 702-19: Reliability, Availability, and Maintainability," February 12, 2020.

Department of the Navy, "NAVSUP Weapons Systems Support Instruction 4441.15L," June 26, 2017.

Dixon, Matthew C., *The Maintenance Costs of Aging Aircraft: Insights from Commercial Aviation*, Santa Monica, Calif.: RAND Corporation, MG-486-AF, 2006. As of September 12, 2022:
https://www.rand.org/pubs/monographs/MG486.html

DoD—*See* U.S. Department of Defense.

Eckstein, Megan, "VNCO: Enthusiasm over Navy's Performance-to-Plan Readiness Effort Is Spreading," *UNSI News*, May 24, 2021. As of May 12, 2022:
https://news.usni.org/2021/05/24/vcno-enthusiasm-over-navys-performance-to-plan-readiness -effort-is-spreading

Ganey, Thomas A., *Endangered Species-Solid Rocket Motor Manufacturers: Preventing a National Asset Extinction*, Air University, Air Command and Staff College, Maxwell Air Force Base, February 11, 2011.

GAO—*See* Government Accountability Office.

Gladstone, Brian, Brandon Gould, and Prashant Patel, "Evaluating Solid Rocket Motor Industrial Base Consolidation Scenarios," IDA Research Notes, Spring 2016. As of May 11, 2022:
https://www.ida.org/-/media/feature/publications/r/rn/rn2016-evalsolidrocketmotor/rn2016 -evalsolidrocketmotor.ashx

Government Accountability Office, "Solid Rocket Motors: DOD and Industry Are Addressing Challenges to Minimize Supply Concerns," GAO-18-45, October 2017. As of May 11, 2022:
https://www.gao.gov/assets/gao-18-45.pdf

———, "F-35 Aircraft Sustainment: DOD Needs to Address Substantial Supply Chain Challenges," GAO-19-321, April 2019. As of May 12, 2022:
https://www.gao.gov/assets/gao-19-321.pdf

———, "Weapon System Sustainment: Aircraft Mission Capable Rates Generally Did Not Meet Goals and Cost of Sustaining Selected Weapon Systems Varied Widely," GAO-21-101SP, November 19, 2020. As of May 12, 2022:
https://www.gao.gov/products/gao-21-101sp

Gunzinger, Mark A., "Affordable Mass: The Need for a Cost-Effective PGM Mix for Great Power Conflict," *Mitchell Institute Policy Paper* 31, November 2021. As of May 11, 2022:
https://mitchellaerospacepower.org/wp-content/uploads/2021/11/Affordable_Mass_Policy _Paper_31-FINAL.pdf

Herrera, G. James, and Brendan W. McGarry, "Defense Primer: Working Capital Funds," Congressional Research Service, updated December 2, 2021. As of May 12, 2022: https://crsreports.congress.gov/product/pdf/IF/IF11233

Hipolite, Whitney, "3D Printed Guided Missiles Are Now a Reality Thanks to Raytheon," *3DPrint.com*, July 16, 2015. As of May 3, 2022: https://3dprint.com/81850/3d-printed-guided-missiles/

Hull, Theresa S., "Audit of Purchases of Ammonium Perchlorate Through Subcontracts with a Single Department of Defense–Approved Domestic Supplier," Office of the Inspector General of the Department of Defense, July 9, 2020.

Janes, "AGM-158C Long-Range Anti-Ship Missile (LRASM) Weapons: Air-Launched," December 15, 2021a. As of May 11, 2022:
Not available to the general public.

———, "Weapons: Naval, Missiles, United States," updated April 7, 2021b. As of May 12, 2022:
https://customer.janes.com/Janes/Display/JNWS0162-JNW_#Status

Joint Chiefs of Staff, "Joint Logistics," Joint Publication 4-0, updated May 8, 2019. As of May 12, 2022:
https://www.jcs.mil/Portals/36/Documents/Doctrine/pubs/jp4_0ch1.pdf

"Joint Direct Attack Munition," *Defense Daily*, 2022. As of April 25, 2022:
https://www.defensedaily.com/joint-direct-attack-munition-jdammanufacturerboeing/

Jones, Brian, "Integrated Supply-Chain Management Pays Off for Naval Supply Systems Command, Partners," U.S. Navy Press Office, November 5, 2020. As of August 18, 2021:
https://www.navy.mil/Press-Office/News-Stories/Article/2406206/integrated-supply-chain-management-pays-off-for-naval-supply-systems-command-pa/

———, "Naval Supply Systems Command Cash War Room Keeping NWCF Solvent," U.S. Navy Press Office, September 17, 2021. As of May 12, 2022:
https://www.navy.mil/Press-Office/News-Stories/Article/2780167/naval-supply-systems-command-cash-war-room-keeping-nwcf-solvent/

———, "NAVSUP WSS Integrated Weapons Support Team Slashes Unfilled GCU Orders to Zero," *DVIDS*, October 29, 2021. As of May 12, 2022:
https://www.dvidshub.net/news/408350/navsup-wss-integrated-weapons-support-team-slashes-unfilled-gcu-orders-zero

Keating, Edward G., Ellen M. Pint, Christina Panis, Michael H. Powell, and Sarah H. Bana, *Defense Working Capital Fund Pricing in the Defense Finance and Accounting Service*, Santa Monica, Calif.; RAND Corporation, RR-866, 2015.

Keller, Jared, "The Pentagon Has Dropped So Many Bombs on ISIS We're Literally Running Out," *Task and Purpose*, May 1, 2017. As of January 10, 2022:
https://taskandpurpose.com/gear-tech/pentagon-bomb-shortage-isis/

Knight-Ridder, "Allies Reportedly Facing Ammunition Shortage. Some Fear Bullets in Short Supply for a Ground War. Persian Gulf Showdown," *Baltimore Sun*, February 13, 1991. As of January 7, 2022:
https://www.baltimoresun.com/news/bs-xpm-1991-02-13-1991044120-story.html

Lambeth, Benjamin S., *NATO's Air War in Kosovo: A Strategic and Operational Assessment*, Santa Monica, Calif.: RAND Corporation, MR-1365, 2001.

———, *Air Power Against Terror: America's Conduct of Operation Enduring Freedom*, Santa Monica, Calif.: RAND Corporation, MG-166-1-CENTAF, 2006. As of April 25, 2022:
https://www.rand.org/pubs/monographs/MG166-1.html

Lynch, Kristin F., Anthony DeCicco, Bart E. Bennett, John G. Drew, Amanda Kadlec, Vikram Kilambi, Kurt Klein, James A. Leftwich, Miriam E. Marlier, Ronald G. McGarvey, Patrick Mills, Theo Milonopoulos, Robert S. Tripp, and Anna Jean Wirth, *Analysis of Global Management of Air Force War Reserve Materiel to Support Operations in Contested and Degraded Environments*, Santa Monica, Calif.: RAND Corporation, RR-3081, 2021. As of May 12, 2022:
https://www.rand.org/pubs/research_reports/RR3081.html

Marceau, James, "Viewpoint: Performance Based Logistics Contracting—Does It Work?" *National Defense*, August 8, 2018. As of May 12, 2022:
https://www.nationaldefensemagazine.org/articles/2018/8/8/viewpoint-performance-based-logistics-contracting-does-it-work

Martin, Bradley, Roland J. Yardley, Phillip Pardue, Brynn Tannehill, Emma Westerman, and Jessica Duke, *An Approach to Life-Cycle Management of Shipboard Equipment*, Santa Monica, Calif.: RAND Corporation, RR-2510, 2018.

O'Connell, Caolionn, Elizabeth Hastings Roer, Rick Eden, Spencer Pfeifer, Yuliya Shokh, Lauren A. Mayer, Jake McKeon, Jared Mondschein, Phillip Carter, Victoria A. Greenfield, and Mark Ashby, *Managing Risk in Globalized Supply Chains*, Santa Monica, Calif.: RAND Corporation, RR-A425-1, 2021.

Office of the Chief of Naval Operations (OPNAV), Navy War Reserve Materiel Program, Office of the Chief of Naval Operations Instruction 4080.11D, January 21, 1999.

OSD A&S Industrial Policy, *Fiscal Year 2020 Industrial Capabilities Report*, January 2021. As of May 12, 2022:
https://media.defense.gov/2021/Jan/14/2002565311/-1/-1/0/FY20-INDUSTRIAL-CAPABILITIES-REPORT.PDF

Persons, Timothy M., "3D Printing: Opportunities, Challenges, and Policy Implications of Additive Manufacturing," *GAO-15-505SP-Addictive Manufacturing Forum*, June 2015. As of June 10, 2022:
https://www.gao.gov/assets/gao-15-505sp.pdf

Plante, Chris, and Charles Bierbauer, "Pentagon's Supply of Favorite Weapon May Be Dwindling," *CNN*, March 30, 1999. As of January 7, 2022:
http://edition.cnn.com/US/9903/30/kosovo.pentagon/

Pyles, Raymond A. and Hyman L. Shulman, *United States Air Force Fighter Support in Operation Desert Storm*, Santa Monica, Calif.: RAND Corporation, MR-468-AF, 1995. As of March 30, 2022:
https://www.rand.org/pubs/monograph_reports/MR468.html

Randol, Devin S., "Commander of NAVSEA Visits MOC AOG," Commander, Naval Air Force Atlantic (AIRLANT), March 4, 2021. As of July 26, 2021:
https://www.airlant.usff.navy.mil/Press-Room/News-Stories/Article/2524520/commander-of-navsea-visits-moc-aog/

Richanbach, Paul H., H. T. Johnson, Kathleen M. Conley, Graeme R. Douglas, Michael F. Fitzsimmons, Kenneth A. Evans, and David R. Graham, *Independent Review of the Defense Logistic Agency's Roles and Missions*, Institute for Defense Analyses, December 2014. As of May 12, 2022:
https://apps.dtic.mil/sti/pdfs/AD1015130.pdf

Robbins, Marc, James R. Broyles, Josh Girardini, Kristin Van Abel, and Patricia Boren, *Improving DoD's Weapon System Support Program*, Santa Monica, Calif.: RAND Corporation, RR-2496-OSD, 2019. As of September 9, 2022:
https://www.rand.org/pubs/research_reports/RR2496.html

Ryder, Dianne, "Rare but Ready," Defense Logistics Agency, December 26, 2016. As of May 12, 2022:
https://www.dla.mil/AboutDLA/News/NewsArticleView/Article/1041913/rare-but-ready/

Saunders, Gregory E., and Nicole Dumm, "Diminishing Manufacturing Sources and Material Shortages," *Defense Standardization Program Journal*, October/December 2017. As of May 12, 2022:
https://www.dsp.dla.mil/Portals/26/Documents/Publications/Journal/171001-DSPJ.pdf?ver=2018-02-06-101522-157

Scott, Richard, "Cruise Control, Block V Missile Starts a New Chapter for Tomahawk Line," *Jane's International Defence Review*, December 9, 2021a. As of May 11, 2022:
https://customer.janes.com/Janes/Display/BSP_10383-IDR

———, "USN Axes JSOW ER in Favour of JASSM-ER Buy," *Jane's Missile and Rockets*, June 9, 2021b. As of May 12, 2022:
https://customer.janes.com/Janes/Display/BSP_481-JMR

———, "Anti-Surface Warfare: USN Plans Funding Towards a Bigger Punch, Longer Reach," *Jane's International Defence Review*, April 29, 2022. As of May 11, 2022:
https://customer.janes.com/Janes/Display/BSP_20376-IDR

Seligman, Lara, "Air Force Wants Smart Bomb Increase for ISIS Fight," *Defense News*, April 1, 2016. As of April 25, 2022:
https://www.defensenews.com/air/2016/04/01/air-force-wants-smart-bomb-increase-for-isis-fight/

Shinkman, Paul D., "ISIS War Drains U.S. Bomb Supply," *U.S. News*, February 17, 2017. As of January 10, 2022:
https://www.usnews.com/news/world/articles/2017-02-17/us-raiding-foreign-weapons-stockpiles-to-support-war-against-the-islamic-state-group

Staff Report Joint Operations Command, "Combat Role in Iraq Complete; Invitation from Iraq Reaffirmed to Advise, Assist, Enable," Operation Inherent Resolve, December 9, 2021. As of April 25, 2022:
https://www.inherentresolve.mil/Releases/News-Releases/Article/2867285/combat-role-in-iraq-complete-invitation-from-iraq-reaffirmed-to-advise-assist-e/

Stahl, Scott CAPT, "Naval Sustainment System—Supply (NSS-S) Navy Working Capital Fund (NWCF) Optimization Pillar," *The Navy Supply Corps Newsletter*, Summer 2021.

Trimble, Steve, "The Weekly Debrief: More Details Emerge About New USAF Mystery Missile," *Aviation Week & Space Technology*, April 5, 2022. As of April 22, 2022:
https://aviationweek.com/defense-space/missile-defense-weapons/weekly-debrief-more-details-emerge-about-new-usaf-mystery

Tripp, Robert S., Kristin F. Lynch, John G. Drew, and Edward W. Chan, *Supporting Air and Space Expeditionary Forces: Lessons from Operation Enduring Freedom*, Santa Monica, Calif.: RAND Corporation, MR-1819-AF, 2004. As of April 22, 2022:
https://www.rand.org/pubs/monograph_reports/MR1819.html

U.S. Code, Title 10, Armed Forces § 2208, Working Capital Funds, 1994.

U.S. Department of the Air Force, *Department of Defense Fiscal Year (FY) 2021 Budget Estimates: Air Force Justification Book, Volume 1, Missile Procurement*, Air Force, February 2020. As of May 11, 2022:
https://www.saffm.hq.af.mil/Portals/84/documents/FY21/PROCUREMENT_/FY21%20Air%20Force%20Missile%20Procurement_1.pdf?ver=2020-02-10-145322-973

U.S. Department of Commerce, Bureau of Industry and Security Office of Technology Evaluation, *U.S. Rocket Propulsion Industrial Base Assessment*, 2018. As of May 11, 2022: https://www.bis.doc.gov/index.php/documents/technology-evaluation/2389-u-s-rocket -propulsion-industry-2018/file

U.S. Department of Defense, *Report to Congress: Kosovo/Operation Allied Force After-Action Report*, January 31, 2000.

———, *SRM Industrial Capabilities Report to Congress: Redacted Version*, June 2009. As of May 11, 2022: https://forum.nasaspaceflight.com/index.php?action=dlattach;topic=18115.0;attach=157687

———, *Selected Acquisition Report—Joint Direct Attack Munition (JDAM)*, December 2018. As of May 13, 2022: https://www.esd.whs.mil/Portals/54/Documents/FOID/Reading%20Room/Selected_Acquisition _Reports/FY_2018_SARS/19-F-1098_DOC_46_JDAM_SAR_Dec_2018.pdf

———, *Advantage at Sea: Prevailing with Integrated All-Domain Naval Power*, December 2020a. As of May 11, 2022: https://media.defense.gov/2020/Dec/16/2002553074/-1/-1/0/TRISERVICESTRATEGY.PDF

———, *Fiscal Year (FY) 2021 Budget Estimates: Navy Justification Book, Volume 1 of 1, Weapons Procurement, Navy*, February 2020b. As of May 11, 2022: https://www.secnav.navy.mil/fmc/fmb/Documents/21pres/WPN_Book.pdf

———, "Operation Inherent Resolve: Targeted Operations to Defeat ISIS," 2022. As of April 25, 2022: https://dod.defense.gov/OIR/

U.S. Department of Defense Inspector General, *Audit of Navy and Defense Logistics Agency Spare Parts for F/A-18 E/F Super Hornets*, DODIG-2020-030, November 19, 2019. As of May 12, 2022: https://media.defense.gov/2019/Nov/21/2002214559/-1/-1/1/DODIG-2020-030.PDF

U.S. Senate, Committee on Armed Services, "Conduct of Operation Enduring Freedom," Hearing before the Committee on Armed Services United States Senate, 107th Cong., 2nd Sess., February 7 and July 31, 2002.

Weisgerber, Marcus, "Bombs Away! Lockheed Expanding Missile Factories, Quadruples Bomb Production for ISIS Long Haul," *Defense One*, March 16, 2016. As of January 10, 2022: https://www.defenseone.com/business/2016/03/lockheed-expands-munitions-factories-isis -future/126725/

Williams, S. W., F. Martina, A. C. Addison, J. Ding, G. Pardal, and P. Colgrove, "Wire + Arc Additive Manufacturing," *Materials Science and Technology*, Vol. 23, No. 7, February 9, 2016. As of June 10, 2022:
https://www.tandfonline.com/doi/full/10.1179/1743284715Y.0000000073

Wolfe, Frank, "Pentagon Speeds Up JDAM Delivery for Possible Kosovo Use," *Defense Daily*, Vol. 202, No. 12, April 16, 1999, p. 1. As of June 10, 2022:
https://www.proquest.com/trade-journals/pentagon-speeds-up-jdam-delivery-possible-kosovo/docview/234093229/se-2?accountid=25333

Younassi, Obaid, Kevin Brancato, John C. Graser, Thomas Light, Rena Rudavsky, and Jerry M. Sollinger, *Ending F-22A Production: Costs and Industrial Base Implications of Alternative Options*, Santa Monica, Calif.: RAND Corporation, MG-797-AF, 2010.